Zoltan J Kiss

Gravitation:
our Quantum Treasure

2013

Order this book online at www.trafford.com
or email orders@trafford.com

Most Trafford titles are also available at major online book retailers.

Printed in the United States of America.

ISBN: 978-1-4907-1073-0 (sc)
ISBN: 978-1-4907-1072-3 (e)

Library of Congress Control Number: 2013914129

Trafford rev. 08/02/2013

 www.trafford.com

North America & international
toll-free: 1 888 232 4444 (USA & Canada)
fax: 812 355 4082

to my friends in London
and
in Bulgaria
to Keith and Rosa Hardwick

Executive Summary

Gravitation has been granted to us.
Gravitation is an energy source of infinite capacity.
Gravitation affords us an opportunity to utilise its *beauty* and *power*!
Gravitation is our quantum treasure!

However, we have never really understood what *gravitation* is about.

The evidence of its real nature is given to us in black and white, in each moment and in every one of our actions. The truth, however, is so strange that it is hard to believe:
Free fall means *Earth's* surface is "falling" under our feet!
   Earth is in sphere symmetrical expanding acceleration at constant speed.
The surface of the *Earth* – together with us on the surface – is in expanding motion,
$$\text{speed of } i = \lim g\Delta t = c.$$

We should use this gift of gravitation, granted to us, rather than continue with the practice of converting the environment into wastes of different kinds. This is not merely to agitate for saving the environment; rather to make the point that causing damage, when we are granted the source of infinite capacity, makes no sense and cannot be our objective.

Energy production is our weak point.
All objects of our life are part of a cycle: what we eat, what we wear, what we use - all are coming from cycles. All our tools, our buildings, all our technology are products of nature, a kind of transformation.
Energy alone is the one by means of which we destroy all around us.
At the same time, *energy* is given to us in the easiest way: in the form of *gravitation*!
This is proven in every single moment of our lives. It is just that we do not take note of it, because it is so obvious and so unbelievable.

The formula is simple: *energy* is granted to us in infinite value!
Earth is in sphere symmetrical expanding acceleration = *gravitation*, functioning as electron process.
Gravitation as quantum treasure is about the use of its quantum impact as *electricity* and *mechanical work*!

To arrive at these conclusions, however, takes us on a long journey.

The first point to realise is: there is no event without time and there is no time without event!

We do not really use the *Albert Einstein* and *Max Planck* legacy.

We like to speak about *Einstein's* time formula and about *Planck's* constant, but without really believing what the message is: *time* is part of our existence; the world and we ourselves are processes in our every single "parts"; *quantum* as such is an impact, intensity, acting in time!

With all respect we shall ask:

Do we understand and, most importantly, do we use in our practice *Einstein's* relativity concept?

Do we in our conventional understanding really consider *world* is with a definite end and with a particular "last elementary component"? Will we go there and will we lounge around?

Do we use *time* and *speed* in our science practice today as decisive key parameters, establishing systems of reference?

Do we use the energy balance concept of relativity?

These concerns relate to all "particle" type definitions and concepts, but the "last particle" *Higgs-Boson* case turns the focus really on it.

Do we fully see what *Max Planck* has given by his constant?

Have we ever wondered why its dimension is *Joule* · sec?

Not really, otherwise we would have come to the point that this is the work value of an *impulse,* acting in *time.*

Zero as such only exists as inflexion point; the checkpoint between positive and negative; the transition point between proton and neutron processes.

And elementary particles are processes.

Life is constant change. The goal is for being in balance – the rule of *Nature.*

The entropy value of elementary processes is the corner stone establishing the *energy/mass quantum.* The correct definition is *energy/mass quantum,* indeed. We cannot decide what part of the mass/energy - energy/mass transformation the quantum entropy belongs to. It should belong to both.

The importance of the *entropy* is that it embodies the principle: there is no perfect transformation. Any change has its price. Quantum is establishing the *Quantum System of Reference* in its accumulating stage, which in its loaded status is the *Quantum Membrane.*

Energy and mass are in fact equal categories, just energy usually has the character of the "result" and mass the "appearance". Proton process, within one and the same elementary transformation, results in energy, neutron process in mass.

The *Quantum System of Reference* has its speed value of quantum communication. The higher the load of the *Quantum Membrane* is, the higher is the speed of its quantum communication.

The speed of light (as we call it in our conventional terminology) is the speed of the propagation of a certain quantum impact. We call this quantum impact "light" within our *Quantum System of Reference* on the surface of *Earth*. But there is no difference whether we see impacts propagating in the quantum system or not (as the case is with radio frequencies), their speed of quantum communication is one and the same.

The speed of quantum communication is not constant, as nothing in *Nature* can be called constant. Its value depends on the load of the *Quantum Membrane*. Within our *Quantum Membrane* on the surface of the *Earth* it has its definite measured value. Within *Earth's plasma* quantum communication is of infinite speed value, the result of infinite conflict and infinite intensity.
The speed value of the quantum impact, coming through the depth levels of the core, the crust and the soil of the *Earth* is getting less and less. The quantum impact of *Earth's plasma* is the drive of *Earth's* sphere symmetrical expanding acceleration = *gravitation*.

Elements are processes, products of the *plasma* conflict. Elements are the building blocks of minerals and the hardened structure of the *Earth*. Each elementary process in its natural mineral stage has its own quantum speed of communication. The deeper the position of the mineral of an element to be found within *Earth* crust is, the higher is its quantum speed value.

The time system of elementary processes is different from our time system on the surface of the *Earth*. This is the reason we measure them as particles. The time system of elements is infinite times slower than ours, the intensity of their processes is infinite times more.

Elements are like our families – cycles of generations: proton, neutron and electron processes in infinite sequence. There should be no surprise and no smile comparing these "components" to real family members: mother as proton, father as neutron and children as electrons. Why would the rules of *Nature* be different for the elements and for us?!
The key of elementary processes is the *blue shift* provided by the electron process, initiating *red shift,* the drive of the neutron collapse. The more the *Quantum Membrane* is loaded, the more intensive the neutron process is. Can we compare this with the structure of the operation of a family? Yes, we easily can.
The electron process is the continuation of the proton process, just in a different time system. The inflexion point is the common point and the change of the neutron and proton processes.
The intensity of the neutron collapse establishes the *inflexion point*, the turning point of the neutron collapse into proton expansion: The intensity of the turnover, the "depth" of the "*zero point*" is the function of the load of the *Quantum Membrane*.
The higher the load is, the higher is the $\pm (dm/dt)$ intensity gradient – the higher is the intensity of the inflexion point. The higher the intensity of the change at the inflexion point is, the more intensive is the unresolved energy intensity reserve (*quantum*) of the mass energy balance and so the load of the *Quantum Membrane*.

If the process is in balance and stable, everybody in the family is happy.
If the balance is destroyed, the isotope stage is full of problems.

Isotopes are the results of the damage of the elementary balance:
Beta is about electron process intensity problems: if it is more than it should be it results in positron radiation, giving off *blue shift* impact; if less, the consequence is electron radiation, taking *blue shift* impact from external elementary sources.
Gamma is serious damage in the elementary balance. The neutron process turns around and is acting as proton would do.
Neutron radiation is the result of the destruction of the elementary balance between proton and neutron processes. The damaged element is in proton process demand and takes proton process cover from other elements.
Alpha radiation is the process of self-correction of the increased intensity of the element. As *Hydrogen* cannot be released through having an infinite long neutron process and so has never been completed, the only option elements with increased energy intensity have is, to release the next in the row element: *Helium*.

If we know the reason of the formation of isotopes, we become also aware what the best ways of their rehabilitation are. Conventional physics, on particle based approach refuses to accept half-life of isotopes can be impacted. But controlled rehabilitation of isotopes is far not a dream or miracle. Elementary processes are events, happening in time and can be influenced.

Hydrogen is one of the countless beauties of *Nature*.
Do we really think that *Nature* has forgotten about the neutron process of the last elementary process within the infinite chain of elementary transformation?
Do we really think that *Nature* creates elements on different basics?
There is a huge difference *between* accepting the fact that we cannot measure the existence of the neutron process in our time system and stating that *Hydrogen* does not have neutron!
Hydrogen has its never-seems-to-be-ending neutron process with infinite low intensity. There is an "end", however, with a certain inflexion feature: the *Big Bang* itself when all cycles and generation processes start again.

Water, *hydro-carbons* are our life components with energy surplus.

The majority of elements are with neutron process dominance.
There are, however, *eight* elements with proton process dominance and electron process *blue shift* surplus: *Hydrogen, Oxygen, Nitrogen, Helium, Carbon, Calcium, Sulphur* and *Silicon*. These are our "saviours": the basic elements of our life.
Aluminium, Magnesium, Potassium (K), Sodium (Na), Chlorine, Phosphorus, Nickel, Fluor, Titanium, Vanadium are with slight neutron process dominance and electron process *blue shift* deficit but very close to equilibrium stage.
Because of this very close-to-the-balance stage status, these elements are very important components for our life in various compositions with the group of the *eight*.

There are proton and neutron processes available within the soil on the surface of the *Earth*, but without the necessary electron process *blue shift* drive. The missing electron process *blue shift* drive is utilised by *gravitation*. In this way, neutron processes have not been driven and proton processes have not been used to their full capacity.

The only *blue shift* drive is coming from those elements surrounding us with *blue shift* surplus: *Hydrogen, Oxygen, Nitrogen, Helium, Carbon, Calcium, Sulphur and Silicon.*

In other words, soil is full of energy, but without the necessary initiation of the process. *Oxygen, Hydrogen, Nitrogen, Carbon* and *water* are the drives and we ourselves and the vegetation around us are the ones establishing quantum contact and making life happen.

Carbon is the element with the most balanced proton-neutron process relation. This is the main reason a diamond as a definite *Carbon* based mineral has its "non-breakable" elementary structure. The speed value of the quantum communication of the *Carbon* element is $c_C = 300,000$ km/sec.

The calculated speed values of the quantum communication of *Hydrogen, Oxygen and Nitrogen* elements are less than the quantum speed (speed of light) on the surface of the *Earth*. *Gravitation* makes them gaseous and results in their quantum speed value of $c = 300,000$ km/sec on the *Earth* surface.

Gold has its quantum speed of communication: $c_G = 366,000$ km/sec

Uranium is with the highest intensity of its electron process *blue shift* impact and quantum communication. Its quantum speed is: $c_U = 378,000$ km/sec.

Why do minerals keep elements in balanced state with different speeds of quantum communication?

The reason is simple: elements represent the resulting balanced status (created at that depth level of *Earth* and at that time point) between the driving *blue shift* conflict of the *plasma* and the expanding and cooling effect of *Earth gravitation*.

These specific statuses have been identified as *different elements*.

Magnetic effect is the acting *Strong Interrelation* across the *Quantum Membrane*. Electron process *blue shift* impact flow (electricity) intensifies proton and neutron processes and this way their interrelation as well. There is no difference does magnetic effect *Strong Interrelation* work within the *Quantum Membrane* of solid elementary structures or is connecting its end surfaces with proton process and neutron process dominances.

Interruption of magnetic lines generates electricity.

Gravitation as quantum treasure is granted to us for use.

Gravitation, the sphere symmetrical expanding acceleration of *Earth,* is the product of the expanding potential of *Earth's* mass/energy internals.

In other words, the increased quantum speed of natural minerals is part of the driving force of *Earth gravitation*. The *blue shift* conflict of the expanding *plasma* status is the engine, but all hardened minerals deep in *Earth* are part of it. Reserves are deep inside.

Gravitation is *blue shift* impact with energy loss. *Earth* is in permanent loss of *blue shift* impact because of *gravitation*.

Rotation and *gravitation* stimulate electron process *blue shift* impact flow within the *Earth*. While *Earth* itself is with proton process dominance in general, the motion of the *blue shift* impact generates magnetic effect between the two ends of the rotation with relativistic proton process dominancy at one end (*North pole*) and neutron process dominancy (*South pole*) at the other.

Our human life and the history of human culture are and have always been about the search and need for finding new energy sources. The classic format of our energy is *electricity*, the propagating electron process *blue shift* impact.

Gravitation is electron process *blue shift* impact!

We have the options for the use of *gravitation* for the benefit of all of us:

- Elementary processes of pulverised mass in free fall *generate electricity*; result of the impact of the interruption of *Earth* magnetic lines.

- Electron process *blue shift* impact acting from a certain height and *blue shifted* in collision by *Earth* surface *generates electricity*; the form is *red shift* at the level of the release of the original *blue shift* impact.

- *Blue shift* conflict generates *Quantum Membrane* of increased intensity above *Earth* surface; *mechanical* (lifting) capability is the benefit for potential use.

Gravitation: our quantum treasure is an energy source of infinite capacity!

Table of Content

Gravitation: our Quantum Treasure
is the fourth book of the concept, elements and *Nature* are processes.

The Energy Balance of Relativity, the first book is about the mass/energy balance, the ruling principle of the communication of different relativistic time systems.
All around us is happening in time, differences are in the intensities of the processes. There are infinite time systems existing simultaneously. The intensity of events, measured as impact within systems of reference is function of the time flow. The same "absolute" event has different impact within the elementary world and within the system of reference of the *Earth*.
Communication between systems of reference, their relation to each other depends on their intensity relations. Systems can communicate with each other if time systems are the same.

The Quantum Energy and Mass Balance, the second book is discussing the process based atomic structure. Energy/mass *quantum* is entropy product, the infinite small mass/energy balance difference between proton and neutron processes. Electron process is the *blue shift* drive of the neutron collapse, the end product of the proton expansion. Electron process is sphere symmetrical expanding acceleration at constant speed.
The relation of proton and neutron processes of elements is established by the intensity of the electron process. The intensity coefficient of the electron process is the distinguishing characteristic of elements. Its value for the known elements of the Periodic Table varies from $\lim \varepsilon_x = 0$ to $\varepsilon_x = 1.58$. *Hydrogen* as unique element has a neutron process of infinite low intensity.

Quantum Engine, the third book, provides further explanations and arguments of the process based elementary approach, with examples and options for the utilisation of the quantum energy.

Reading the book without the background of the first three books, *the most important definitions* are:

Quantum System of Reference	accumulation of mass/energy quantum within certain defined quantum space
Quantum Space = space	the space, where events happen in time
Quantum Membrane	the loaded status of the *Quantum System of Reference*; the load is electron process *blue shift* impact
Proton process	sphere symmetrical expanding acceleration from $v = 0$ (the inflexion point) to $i = \lim v = c$
Neutron process	sphere symmetrical accelerating collapse from $i = \lim v = c$ to $v = 0$ (the inflexion point)
Electron process	sphere symmetrical expanding acceleration at constant $i = \lim a\Delta t = c$ speed

The most important formulas are:

Speed accelerates time flow: $dt_v = \dfrac{dt_o}{\sqrt{1 - \dfrac{v^2}{c^2}}}$

The intensity of the *proton process*

$$e_p = \frac{dmc_x^2}{dt_p}\left(1 - \sqrt{1 - \frac{v^2}{c_x^2}}\right) = \frac{dmc_x^2}{dt_o}\sqrt{1 - \frac{v^2}{c_x^2}}\left(1 - \sqrt{1 - \frac{v^2}{c_x^2}}\right);$$

where $0 \Rightarrow v \Rightarrow i_x$

     and at the end-stage: $e_p = \dfrac{dmc_x^2}{dt_o}\sqrt{1 - \dfrac{i_x^2}{c_x^2}}\left(1 - \sqrt{1 - \dfrac{i_x^2}{c_x^2}}\right) = \dfrac{dmc_x^2}{dt_{ix}}\left(1 - \sqrt{1 - \dfrac{i_x^2}{c_x^2}}\right);$

The intensity of the *neutron process*: $e_n =$

$$= \frac{dmc_x^2}{dt_{ix}}\sqrt{1 - \frac{(c_x - i_x)^2}{c_x^2}}\left(\frac{1}{\sqrt{1 - \dfrac{u^2}{c_x^2}}} - 1\right) = \frac{dmc_x^2}{dt_o}\sqrt{1 - \frac{i_x^2}{c_x^2}}\sqrt{1 - \frac{(c_x - i_x)^2}{c_x^2}}\left(\frac{1}{\sqrt{1 - \dfrac{(i_x - v)^2}{c_x^2}}} - 1\right)$$

         where $i_x \Rightarrow v \Rightarrow 0$

at the end-stage:

$$e_n = \frac{dmc_x^2}{dt_o}\sqrt{1 - \frac{i_x^2}{c_x^2}}\sqrt{1 - \frac{(c_x - i_x)^2}{c_x^2}}\left(\frac{1}{\sqrt{1 - \dfrac{i_x^2}{c_x^2}}} - 1\right) = -\frac{dmc_x^2}{dt_o}\sqrt{1 - \frac{(c_x - i_x)^2}{c_x^2}}\left(1 - \sqrt{1 - \frac{i_x^2}{c_x^2}}\right)$$

Proton start stage: $\dfrac{dmc_x^2}{dt_o}$;   Proton end stage: $\dfrac{dmc_x^2}{dt_o}\sqrt{1 - \dfrac{i_x^2}{c_x^2}}$;

Neutron start stage: Neutron end stage:

$\dfrac{dmc_x^2}{dt_o}\sqrt{1 - \dfrac{i_x^2}{c_x^2}}\sqrt{1 - \dfrac{(c_x - i_x)^2}{c_x^2}}$; $\dfrac{dmc_x^2}{dt_o}\sqrt{1 - \dfrac{(c_x - i_x)^2}{c_x^2}}$;

The intensity of the electron process: $e_e = \dfrac{dmc_x^2}{dt_{ix}\varepsilon_{ix}}\left(1 - \sqrt{1 - \dfrac{(c_x - i_x)^2}{c_x^2}}\right)$

Intensity coefficient of the electron process: $\varepsilon_{ix} = \dfrac{\varepsilon_p}{\varepsilon_n}\sqrt{1 - \dfrac{(c_x - i_x)^2}{c_x^2}}$;

Absolute balance of elementary processes:

$$\frac{dmc^2}{dt_p \varepsilon_p}\left(1 - \sqrt{1 - \frac{i^2}{c^2}}\right) = \frac{dmc^2}{dt_n \varepsilon_n}\, \xi \sqrt{1 - \frac{(c-i)^2}{c^2}}\left(\sqrt{1 - \frac{i^2}{c^2}} - 1\right)$$

ξ representing the entropy consequence of the cycle.

The energy and mass balance of the most important elements of the Periodic Table
(page 101 of Book 2)

Element	Periodic Number P	M Atomic Weight measured	P Proton Mass	N Neutron Mass	$Z = \varepsilon_x$ Event Concentration of the Element, equal to the Intensity Coefficient of the Electron Process
Hydrogen	1	1.0079	1.0072	0.00008	**0.000081**
Helium	2	4.0026	2.0145	1.9869	**0.9863**
Carbon	6	12.0110	6.0436	5.9640	**0.9868**
Nitrogen	7	14.0067	7.0508	6.9519	**0.9860**
Oxygen	8	15.9990	8.0581	7.9364	**0.9849**
Sodium (Na)	11	22.9890	11.0799	11.9030	**1.0743**
Magnesium	12	24.3050	12.0872	12.2111	**1.0102**
Aluminium	13	26.9815	13.0945	13.8798	**1.0560**
Silicon	14	28.0855	14.1017	13.9760	**0.9911**
Phosphorus	15	30.9737	15.1090	15.8564	**1.0495**
Sulphur	16	32.0600	16.1163	15.9349	**0.9887**
Chlorine	17	35.4530	17.1235	18.3200	**1.0699**
Potassium (K)	19	39.0980	19.1381	19.9494	**1.0424**
Calcium	20	40.0800	20.1454	19.9236	**0.9890**
Titanium	22	47.9000	22.1599	25.7280	**1.1610**
Iron	26	55.8470	26.1890	29.6437	**1.1319**
Cobalt	27	58.9332	27.1962	31.7221	**1.1664**
Nickel	28	58.7100	28.2035	30.4911	**1.0811**
Cuprum	29	63.5400	29.2108	34.3132	**1.1746**
Zinc	30	65.3800	30.2181	35.1454	**1.1631**
Silver	47	107.8680	47.3416	60.5005	**1.2779**
Gold	79	196.9665	79.5743	117.3488	**1.4747**
Mercury	80	200.5900	80.5816	119.9645	**1.4887**
Thallium	81	204.3700	81.5888	122.7367	**1.5043**
Lead	82	207.8000	82.5961	125.1589	**1.5153**
Uranium	92	238.0290	92.6688	145.3097	**1.5680**
Plutonium	94	*244*	94.6833	149.2651	**1.5765**

1
World **is infinite**
Zero **and** *"End"* **have specific meaning**

$\mathbf{W}$e deal in our everyday practice with a certain range of data and tend to decide on their significance or non-significance, based on the measurements and changes we experience. But what we measure in our – even most advanced – laboratories, or write into – our most sophisticated – formulas, represent only section of the real world and the content of our findings depends on our capabilities. Categories *zero* (beginning) and *end* as such in fact does not exist in the *universe*.

When we say that our world has *no ends* and when we prove that there is no limit in the directions neither of small nor of large dimensions, one can argue that *c*, the speed of light limits us anyway.

When we are convinced that *status of rest* as such does not exist, one can argue what is the status in this case of the *inflexion point*, where the sphere symmetrical accelerating collapse turns into sphere symmetrical expanding acceleration?

The speed of light in fact is not "the speed of light", rather the *speed* of the transmission of the *quantum impact* of elementary processes by the *Quantum Membrane,* the speed of quantum communication.

The physical value of the speed of the quantum communication, in other words the speed value of the moving quantum impact depends on many factors, first of all the motion and the load of the *Quantum Membrane* of the system of reference of the measurement.

Equal quantum impact of *dn* results in various effects, depending on the time flow (motion) and the intensity of the system of reference of the measurement:

$$\frac{dn}{dt_o} \neq \frac{dn}{dt_v} ; \quad \text{while in absolute values however} \quad \frac{dn}{dt_o \varepsilon_o} = \frac{dn}{dt_v \varepsilon_v}$$

[ε is the intensity coefficient, in general terms reciprocal to the time flow (duration) of events. In the case of the electron process and the proton and neutron transformation, ε is usually more complex than being just simple reciprocal component to time.]

Electron process is an impact to the *Quantum System of Reference*, turning it into *Quantum Membrane*.

If we suppose that number of electron process *blue shift* impacts simultaneously affect a limited size of *Quantum System of Reference* – which is a quite hypothetical, since the *Quantum System of Reference* is of infinite number of energy quantum – the quantum effect, measured within the system of reference of sphere symmetrical expanding acceleration for infinite time is:

1A1
$$\frac{dmc^2}{dt_i \varepsilon_x}\left(1-\sqrt{1-\frac{(c-i)^2}{c^2}}\right) + \frac{dmc^2}{dt_i \varepsilon_y}\left(1-\sqrt{1-\frac{(c-i)^2}{c^2}}\right) + \ldots$$

$$\ldots + \frac{dmc^2}{dt_i \varepsilon_x}\left(1-\sqrt{1-\frac{(c-i)^2}{c^2}}\right) = X\frac{dmc^2}{dt_i \varepsilon_{av}}\left(1-\sqrt{1-\frac{(c-i)^2}{c^2}}\right)$$

1A2   If we suppose that $\dfrac{dmc^2}{dt_i \varepsilon_{av}}\left(1-\sqrt{1-\dfrac{(c-i)^2}{c^2}}\right) = \dfrac{dn}{dt_i \varepsilon_{av}}q$

obviously $X\dfrac{dmc^2}{dt_i \varepsilon_{av}}\left(1-\sqrt{1-\dfrac{(c-i)^2}{c^2}}\right) = X\dfrac{dn}{dt_i \varepsilon_{av}}q$

ε_{av} - means the taken average intensity of the X simultaneous *blue shift* impacts

dXn would be incorrect since we suppose that the impacted part of the *Quantum System of Reference* is limited and the same. So, the number of the impacted energy quantum also should be the same. It is better to express this via the intensity of the quantum impact at the side of the *Quantum Membrane*:

1A3
$$X\frac{dmc^2}{dt_i \varepsilon_{av}}\left(1-\sqrt{1-\frac{(c-i)^2}{c^2}}\right) = \frac{dn}{dt_i(\varepsilon_{av}/X)}q$$

1A4
$$X\frac{dmc^2}{dt_i \varepsilon_{av}}\left(1-\sqrt{1-\frac{(c-i)^2}{c^2}}\right) = \frac{dn}{dt_i \varepsilon_{avX}}q \qquad \text{and} \qquad \varepsilon_{avX} = \frac{\varepsilon_{av}}{X}$$

1A5
$$\text{and obviously } \frac{dn}{dt_i \varepsilon_{avX}} > \frac{dn}{dt_i \varepsilon_{av}}$$

If we would like to express the summarised impact (the summarised load) to the *Quantum Membrane*, the description would be:

1A6
$$\frac{dn}{dt_i \varepsilon_{avX}}q = \frac{dmC^2}{dt_I}\left(1-\sqrt{1-\frac{(C-I)^2}{C^2}}\right)$$

We consider making correction only in c. But $(c-i)$ automatically modifies i as well. Electron process is always at speed $i = \lim v = c$, therefore with the increase of the speed of quantum communication the constant speed of the electron process increases as well.

1A7 $C > c$ and since $I = \lim a\Delta t = C$ as consequence $I > i$

As conclusion we can state that permanent and increased quantum impact result in the increase of the speed of the transfer *(c)* of the impact (signal)! *c*, the speed of quantum communication varies.

At the same time the actual speed of the acceleration higher and therefore the time system of the measurement becomes shorter, meaning more intensive.

dm is taken for standard one, since this is the measured parameter; and $\lim(C - I) = \lim(c - i) = 0$ also can be taken as standard.   1A8

In other words, *c*, the speed of quantum communication ("in conventional terms called as the speed of light") is not an absolute value. It is established, in fact, by the load of the *Quantum Membrane*.

The key of elementary processes is the *blue shift* provided by the electron process, initiating *red shift,* the drive of the neutron process. The more the *Quantum Membrane* is loaded, the more intensive the neutron collapse is.

The intensity of the neutron collapse establishes the *inflexion point*, the turning point of the neutron collapse into proton expansion: The intensity of the turnover and the "depth" of the "*zero point*" (= the slow-down of the collapse) is function of the load of the *Quantum Membrane*.

The higher is the load, the higher is the $\pm(dm/dt)$ intensity gradient the higher is the intensity of the change at the inflexion point. And most importantly, the more intensive is the unresolved energy intensity reserve (*energy quantum*) of the mass energy balance:

$$\frac{dmc^2}{dt_o}\sqrt{1 - \frac{(c-i)^2}{c^2}}\ ; \quad \text{with} \quad \lim dt_o = 0$$ 1B1

As *c* is part of this balance, so is dt_o the time count at the inflexion point.

With reference to 1A8, the value within the square root is the same for all systems in motion, as $\lim(c - i) = 0$

While the *Quantum System of Reference* is common whatever are the dimensions and the time count of systems of reference, *Quantum Membranes* relates to certain cycles "working" in balance. Therefore there might be infinite number of *Quantum Membranes* within the *Quantum System of Reference*. Each *Quantum Membrane* however is in contact and they have impact on each other, as far as the impact is affecting.

    [On the level of quarks we may call it as *Gluon Membrane* - the meaning is the same.]

As $\lim i = c$ and this way: $\lim\dfrac{i}{c} = \lim\dfrac{i_x}{c_x} = \lim\dfrac{I}{C} = 1$ 1B2

and dt_o is the inflexion point for any processes, the time system of electron process communication for all elementary processes is one and the same indeed:

$$dt_i = \frac{dt_o}{\sqrt{1 - \dfrac{i_x^2}{c_x^2}}}$$ 1B3

The points to remember are:

1. The speed of the mass-energy-mass...transformation and the transmission of quantum impacts (as the *blue shift* is) by the *Quantum Membrane,* the speed of *quantum communication* is *c.*

In our *Earth* circumstances the measured $c \cong 300,000,000$ *m/sec* value - alongside the natural conditions of the elementary world – is consequence of gravitation (the *blue shift* impact) of the *Earth* and the loaded status of the *Quantum Membrane* on the surface of the *Earth.* (The last is subject to increasing impact from the activity of the civilisation.)

The speed value of quantum communication varies depending on the load of the *Quantum Membrane* of the system of reference.

2. *Quantum entropy* status is part of the elementary proton-electron-neutron process.

If the intensity of the mass change is $\dfrac{dm}{dt}$ at the beginning of the extension,

$$\text{it is } \quad \frac{dm}{dt}\sqrt{1 - \frac{(c-i)^2}{c^2}} \quad \text{at the end of the collapse.}$$

There is a certain accumulation of the "lost" energy intensity of the elementary cycle – in line with the *entropy* law.

3. The *entropy* of the elementary mass-energy transformation results in generation of energy *quantum,* composing the *Quantum System of Reference.* With reference to 1B3 and 1B4, the intensity of the re-transformation of energy into mass is less at the end than the transformation of mass into energy, as it was at the beginning. If the intensities of the mass change in the two directions are different, there is an accumulation of the energy intensity reserve, composing the *Quantum System of Reference.*

4. Mass-energy and energy-mass transformations in elementary processes (between neutron-proton and proton-neutron processes) happen in parallel, composing the *Strong Interrelation.* The drive of the *Strong Interrelation* is the electron process the *Week Interrelation,* driving the neutron collapse. The *Quantum System of Reference* under electron process *blue shift* impact is the *Quantum Membrane.* The absolute values of the energy/mass transformation of the *Strong Interrelation* are strongly equal, while the intensities of the proton and neutron processes are different.

5. The *Quantum Membrane* is establishing the inflexion point, the point of the neutron-proton process turnover.

Nature establishes the intensity of the mass-energy transformation of elements. The energy-mass re-transformation, result of the load of the *Quantum Membrane,* establishes the intensity of the mass-energy transformation of the next cycle.

In other words:

the impacting natural *blue shift* will result in a consolidated collapse and will result in intensity of the new mass change not higher than the intensity of the collapse was. Since the collapse is driven by *blue shift*.

6. Any external (out of the natural cycle electron process) impact modifies the load of the *Quantum Membrane* and with that the full cycle:

The load impact influences the intensity of the inflexion point and modifies the potential of the unresolved intensity reserve of the *Quantum System of Reference*, the *Quantum Membrane*.

It clearly shown within the formulas, since

$$\sum \frac{dmc^2}{dt}\sqrt{1-\frac{i^2}{c^2}}\left(1-\sqrt{1-\frac{(c-i)^2}{c^2}}\right) = X\frac{dn}{dt_i}q = \frac{dmc^2}{dt_i(1/X)}\left(1-\sqrt{1-\frac{(c-i)^2}{c^2}}\right) \qquad \text{1C1}$$

the summarised or increased electron process *blue shift* impact results in increased intensity of the energy-mass re-transformation (neutron collapse):

$$\frac{dmc^2}{dt_i\frac{1}{X}\varepsilon_X}\sqrt{1-\frac{(c-i)^2}{c^2}} = \frac{dmc^2}{dt_i\varepsilon_i}\sqrt{1-\frac{(c-i)^2}{c^2}} \; ; \quad \text{and} \;\; \varepsilon_X = \varepsilon_i \cdot X \qquad \text{1C2}$$

7. If the external impact is permanent, while the natural correction is effecting, the constant external electron process impact further deepens the neutron collapse, since the additional electron process as per 1D5 provides "external" load: the speed of quantum communication will be of increased value!

As consequence of permanent external impact, the *red shift* will result in $u = i$, which will be equivalent to deep collapse intensity and the mass-energy transformation shall start with high intensity, corresponding to the $v = adt(= c)$ – higher speed of quantum communication.

8. Unresolved energy intensities within the *Quantum System of Reference* and external impacts load the *Quantum Membrane*.

9. Electron process communication is universal, independently on the speed value of quantum communication and the type of the element.

10. Inflexion point and the definition of dt_o as status of *rest* would be incorrect. Inflexion point as such is only acceptable as a certain infinite short status point of the mass-energy transformation of elementary processes.

2
Gravitation =
sphere symmetrical expanding acceleration of the Erath

E_{arth} gravitation is the sphere symmetrical expanding acceleration of
Earth with $a = g = 9.81...m/s^2$
Gravitation is *blue shift* impact, generated by the expansion of our planet.

The time system of the human system on the surface *Earth* is: dt_{HB}
Elementary processes at the surface (and below the surface) of the *Earth* have a time system communicating with the time system of the *Earth*: dt_x
The time system of *Earth* in sphere symmetrical expanding acceleration, the electron function is: dt_E

We on the surface of *Earth* do not feel this acceleration as the time system of the acceleration is infinite long.
We have been part of this motion. Our *time system of reference* (our living circumstances) – corresponds to this motion, acceleration for infinite time, speed of $i_E = \lim g\Delta t = c_E$. At the surface of the *Earth*: $dt_{HB} = dt_E$

2A1   The time system of *Earth*
in line with the sphere symmetrical expanding
acceleration of the *Earth* is
$$dt_E = \frac{dt_o}{\sqrt{1-(i_E/c_E)}}$$

c_E is taken as speed value,   while
2A2   measured on the surface of the *Earth* relative to
the status of the inflexion point
$$dt_o = \frac{dt_o}{\sqrt{1-(0/c_E)}} = dt_o$$

The driving force of the electron processes is coming from the internal conflict of the $i = \lim a\Delta t = c$ speed expansion.
The best characterisation of the status of this internal conflict is: *plasma*. All components (on the surface) of the sphere symmetrical expanding acceleration against the *Quantum Membrane* have $i = \lim a\Delta t = c$ speed value.
Do proton and neutron processes have internal *plasma* conflict?
The answer is easy: No.
The transition of the neutron collapse into proton expansion is the inflexion point.
Neutron collapse is driven by electron process and the necessary energy

cover is provided by the proton process. Therefore there is nothing to be in conflict in neither of the proton or neutron processes.

The end stage of the proton process is the expansion by $i = \lim a\Delta t = c$. This is the start of the conflict of the electron process.

Plasma characterises the electron process *blue shift* conflict of infinite intensity.

Gravitation is *blue shift* load.

Where the *blue shift* impact of *gravitation* is getting generated?

At the level of the *plasma* conflict.

Gravitation, the *blue shift* impact of *Earth* starts not at the surface, as surface is conflicting with the "atmosphere-system of reference", but below it at *plasma* depth level. All around and "above it" are elementary relations and *Quantum System of Reference*.

The speed of the motion of the *Earth's* surface is equal to $i_E = \lim a\Delta t = c_E$

The motion of the *Earth's* surface and with the surface all minerals down to plasma level have this speed value in general. The acceleration is slightly changing from level to level in depth in line with the radius of the planet.

- dt_E relates to c_E and the time system represents the motion of the *Earth* surface with i_E.

- elementary processes on the surface and below the surface of the *Earth* have dt_x the time system of their quantum communication, relating to the intensity of the elementary process.

The speed of quantum communication, measured at the surface of *Earth* is $c_L \approx 300,000$ *km/sec*.

The speed value of *Earth* expanding acceleration is $i_E = \lim a\Delta t = c_E$.

The question is, might these two quantum communication speed values equal and the same?

Yes they are: $c_E = c_L$

The proof is that the quantum impact of light is propagating in accordance with the intensity of the *Quantum Membrane* loaded by the *blue shift* impact of the *Earth* acceleration – *gravitation*.

The communicating dt_E time system on the surface of *Earth* integrates all impacts and effects of elementary processes within the *Earth*, with integrated $c_E \approx 300,000$ *km/sec* speed value of quantum communication.

With reference to 2A1, the time system on the surface of the *Earth* is speeded up infinite times in relation to the inflexion point.

Ref 2A1

Gravitation corresponds to the time system and to the speed of the sphere symmetrical expanding acceleration of the surface of the *Earth*.

2A2
$$e_E = \frac{dmc_E^2}{dt_E}\left(1 - \sqrt{1 - \frac{(c_E - i_E)^2}{c_E^2}}\right) = \frac{dmc_E^2}{dt_o}\sqrt{1 - \frac{i_E^2}{c_E^2}}\left(1 - \sqrt{1 - \frac{(g\Delta t)^2}{c_E^2}}\right)$$

where g is permanently changing and getting always less and less, while

2A3
$$(g\Delta t) = (c_E - i_E) = const \text{ and } \lim(c_E - i_E) = 0$$

and c_E is measured relative to the inflexion point!

Earth is expanding in quantum space. *Earth* is in sphere symmetrical expanding acceleration – de facto – is an electron function.

There is an obvious and measured *blue shift* generation and clear acceleration by g at constant speed value of $i_E = \lim g\Delta t = c_E$.

The expansion and transformation of the mass of *Earth* into *blue shift* impact are parallel processes. This *blue shift* is impacting all elementary and other processes, part of the local *Quantum System of Reference*, the *Quantum Membrane*.

The soil on the *Earth* surface is a specific mass/energy composition with traces and quasi traces of minerals and elementary processes. Quasi traces in general mean no info or no data of the composition.

Ref.
S.11

Soil is certain composition of elementary components, to be identified later in Section 11.

If we ask why soil is taking electron *blue shift* impact (electricity) in infinite volumes, the logic of the answer obliges us to say: because *Earth* is losing electron *blue shift* in infinite volumes.

How *Earth* is losing on *blue shift* impact? The sphere symmetrical expanding acceleration at quasi constant $i = \lim a\Delta t = c$ speed for infinite time is electron function, work against the *Quantum Membrane*. The drive is *blue shift* impact. *Earth* is accelerating and this way is permanently losing its *blue shift* mass/energy potential through its surface.

This energy/mass loss is establishing soil structure.

Ref.
S.9.5

Soil is with quantum communication and *Quantum Membrane*.

With reference to Section 9.5, *plasma* is *blue shift* conflict of infinite intensity. This *blue shift* conflict is the source of *Earth gravitation*, its sphere symmetrical expanding acceleration.

Increased *blue shift* conflict within the depths of the soil means increased speed of quantum communication. Soil is communicating with this speed.

The conflict is less and less towards the surface, which means, acceleration takes the energy of the *plasma blue shift* conflict.

Cooling means losing on electron process *blue shift* impact.

Approaching the surface on *Earth* elementary components of minerals within the soil are losing on *blue shift* conflict, in fact on the speed of quantum communication.

The speed of quantum communication is the only parameter soil components can lose.

$$\frac{dmc^2}{dt_i \varepsilon_x}\left(1-\sqrt{1-\frac{(c-i)^2}{c^2}}\right)$$ 2B1

Losing on the value of c does not mean change of the element within the soil. The closer is the mineral to the surface the less is – with reference to Table 9.1 of Section 9.5 – the speed value of its quantum communication, but still above $c = 300000$ km/sec, the quantum speed of *gravitation*.

Ref. Tab. 9.5

This is the reason of the structure and the consistency of the soil on *Earth* surface. Elementary components are with increased speed of quantum communication and these speed values may differ from each other as approaching *Earth* surface.
Blue shift surplus causes conflict, representing gaseous or liquid states.

Earth soil is solid. Depending first of all on the quality of the composition, but powder like in general at the surface and stone structure format in depths.
Soil is called monatomic in conventional terms in many cases, because of the difficulties for establishing its composition. Soil is about elementary proton/neutron communication just the speed of this communication in the case of deep minerals is higher than our quantum speed of *gravitation*.

The electron process of elements, gravitation and human time systems are communicating. (This is proven by the fact that we experience various forms of electron *blue shift* conflicts on the *Earth*: light, heat, cold, pressure.)
The intensity of our human receipt is:

where $h = 6.63 \cdot 10^{-36}$ *Joulesec*
the *Planck'* constant

$$e_E = \frac{dn}{dt_E}q = n \cdot f \cdot h$$ 2C1

The experienced intensities of the communicating systems are:

Electron process *blue shift* impact of element x is:

$$e_{ix} = \frac{dmc_x^2}{dt_o}\sqrt{1-\frac{i_x^2}{c_x^2}\left(1-\sqrt{1-\frac{(c_x-i_x)^2}{c_x^2}}\right)}$$ 2C2

The time system of the electron process of element x:

$$dt_{ix} = \frac{dt_o}{\sqrt{1-(i_x^2/c_x^2)}}$$ 2C3

The time system at the surface of the *Earth* in communication with the elementary world.

$$dt_E = \frac{dt_o}{\sqrt{1-(i_E^2/c_E^2)}}$$ 2C4

Elementary *blue shift* impacts communicate with the *blue shift* impact of *Earth's* gravitation.
With reference to 2A2 and 2B2 the relation is:

$$\rho = \frac{e_E}{e_{ix}} = \frac{\sqrt{c_E^2-i_E^2}}{\sqrt{c_x^2-i_x^2}} \cdot \frac{c_E-\sqrt{2c_E i_E-i_E^2}}{c_x-\sqrt{2c_x i_x-i_x^2}}$$ 2C5

While the surface of the *Earth* is at speed $i_E = \lim g\Delta t = c_E$ elements have their own cycle with c_x the speed of quantum communication and electron process of $i_x = \lim a_x \Delta t_x = c_x$.

The quantum entropy gradient of elementary processes therefore might be different than that is of the *Earth* with our time system part of it.

2D1   The quantum entropy gradient of the *Earth* is value of:
$$qe_E = \frac{dmc^2}{dt_o}\sqrt{1 - \frac{(c_E - i_E)^2}{c_E^2}}$$

2D2   The quantum entropy gradient of elementary processes is equal to:
$$qe_x = \frac{dmc_x^2}{dt_o}\sqrt{1 - \frac{(c_x - i_x)^2}{c_x^2}}$$

The inflexion point is one and the same for any time systems including *Earth* and elements.

The relation of quantum entropy gradients is function of the speed of quantum communication:

2D3
$$\frac{qe_x}{qe_E} = \frac{c_x^2}{c_E^2}; \quad c_x = c_E\sqrt{\frac{qe_x}{qe_E}}; \quad \text{and} \quad qe_x = qe_E\frac{c_x^2}{c_E^2}$$

Proposing that $\sqrt{1 - \frac{(c_E - i_E)^2}{c_E^2}} \cong \sqrt{1 - \frac{(c_x - i_x)^2}{c_x^2}}$

Higher speed of quantum communication corresponds to higher quantum entropy gradient.

In summary:
The sphere symmetrical expanding acceleration of *Earth* is taking place in a quantum system of reference with time system, different than ours on the surface of the *Earth*.

The common point of these systems of references is the *inflexion point*. The intensity of the *inflexion point* corresponds to the specifics of the system of reference in acceleration.

The sphere symmetrical expanding acceleration of *Earth,* is of quasi constant speed, equal to $i_E = \lim a\Delta t = c_E$.
We live on the surface of the *Earth*, consequently, our system of reference is speeded up by $i_E = \lim a\Delta t = c_E$ relative to the *inflexion point*.

Gravitation is with certain speed of quantum communication as the acceleration is of constant speed, value of $i_E = \lim g\Delta t = c_E$.
The speed of quantum communication of *Earth* is given by and is result of the sphere symmetrical expanding acceleration!

3
Hydrogen, Oxygen, Water and Fire

The neutron process within the elementary structure of the *Hydrogen* is of infinite low intensity. But this is not the neutron process, which establishes the intensity of the collapse. Neutrons are the passive part.

The intensity of the neutron collapse is established by the intensity of the electron process.

The intensity coefficient of the electron process is coming from the relation of $\varepsilon_x = \dfrac{\varepsilon_p}{\varepsilon_n}\sqrt{1 - \dfrac{(c-i)^2}{c^2}}$
3A1

In absolute terms the proton and neutron processes are equal:

$$\frac{dmc^2}{dt_p \varepsilon_p}\left(1 - \sqrt{1 - \frac{i^2}{c^2}}\right) = \frac{dmc^2}{dt_n \varepsilon_n}\xi\sqrt{1 - \frac{(c-i)^2}{c^2}}\left(\sqrt{1 - \frac{i^2}{c^2}} - 1\right)$$
3A2

$$\text{and } \varepsilon_x = \frac{dt_n}{dt_p} = \frac{\varepsilon_p}{\varepsilon_n}\sqrt{1 - \frac{(c-i)^2}{c^2}}$$
3A3

This means the relation of the intensities of the proton and neutron processes establishes the intensity of the electron process *blue shift* impact.

The neutron process is driven by certain intensity value and covered by the proton process.

The intensity coefficient of the electron process of the *Hydrogen* can be written this way:

$$\lim \varepsilon_{xH} = \frac{dt_{n(\to \infty)}}{dt_p} = \lim \frac{\varepsilon_{pH}}{\varepsilon_{nH(\to 0)}}\sqrt{1 - \frac{(c-i)^2}{c^2}} = \infty$$
3A4

In the structure of the *Hydrogen*: $\lim dt_{nH} = \infty$ and $\lim \varepsilon_{nH} = 0$.
3A5

(Later no prefix will be marked in this section, as all relations belong to the *Hydrogen*.)

If $\varepsilon_{pH} > \varepsilon_{nH}$ indeed, 3A2 can be written this way:

$$\frac{dmc^2}{dt_p \varepsilon_p}\left(1 - \sqrt{1 - \frac{i^2}{c^2}}\right) = \frac{dmc^2}{ndt_p \dfrac{\varepsilon_p}{n}}\xi\sqrt{1 - \frac{(c-i)^2}{c^2}}\left(\sqrt{1 - \frac{i^2}{c^2}} - 1\right)$$
3B1

The descriptions in 3B1 and 3A2 are only correct if they are adjusted by ξ, representing the entropy consequence of the cycle. (In intensity format this coefficient is not necessary, as time and intensity values cover it.)

It is supposed (and later it also will be proven) that the proton process of the *Hydrogen* (as end-product of the completed *Helium* cycle) is of "normal-usual" intensity and duration. The neutron process however is of significant nature, consequence of the infinite high intensity coefficient and the infinite low intensity value of the electron process *blue shift* drive.

As being the last element, the neutron process of the *Hydrogen* is of infinite length in time and of infinite low intensity. With reference to 1B1, it can be written in intensity terms like this:

3B2
$$\frac{dmc^2}{dt_p}\left(1-\sqrt{1-\frac{i^2}{c^2}}\right) \neq \frac{dmc^2}{ndt_p}\sqrt{1-\frac{(c-i)^2}{c^2}}\left(\sqrt{1-\frac{i^2}{c^2}}-1\right)$$

Ref 3B1   The proton and neutron processes of the *Hydrogen* are far not equal in intensity terms: $dt_n = n \cdot dt_p$

While the number of the generation of the proton (and electron) processes are of $\lim n = \infty$, the completion of the neutron process is $\lim n_n = 0$, approaching zero: There are infinite numbers of electron processes, as generated from the infinite numbers of proton processes, while there is none of the neutron processes has been completed!

Ref 3A4   With reference to 3A4 the intensity coefficient of the electron process is infinite big value.

It must be noted that the intensity coefficient of the electron process is not the acting intensity of the electron process. It is equal to the proton/neutron intensity relations. The higher is the intensity coefficient the lower is the acting *blue shift* intensity of the electron process. (The neutron process intensity – as result – is in the denominator!)

3B2 means

3B3 $\varepsilon_x = \dfrac{dt_n}{dt_p} = \dfrac{ndt_p}{dt_p}\dfrac{1}{\xi} = \dfrac{n\varepsilon_p}{\varepsilon_p}\sqrt{1-\dfrac{(c-i)^2}{c^2}}$; and $\varepsilon_x = n\sqrt{1-\dfrac{(c-i)^2}{c^2}}$; and $\varepsilon_x = \dfrac{n}{\xi}$

3B4   In 3B3: $\lim\sqrt{1-\dfrac{(c-i)^2}{c^2}}=1$ as $\lim(c-i)=0$ and $\lim i = c$

And 3B3 means:

there are *n* number of electron process *blue shifts*, intensity of infinite *low* value, driving the neutron collapse of the *Hydrogen*.

In accordance with this *blue shift* intensity, the duration of *each* of the neutron processes of the *Hydrogen* is of *infinite* length of duration. There are infinite numbers of neutron processes in the *Hydrogen*, each and all with infinite length of duration.

Hydrogen in natural conditions is in gaseous state. It cannot be differently since the intensity of the neutron process is of infinite low value. This is consequence of the acting intensity of the electron process *blue shift* drive, which is of infinite low value as well:

$$\lim e_e = \lim \frac{dmc^2}{dt_i \varepsilon_x} \left(1 - \sqrt{1 - \frac{(c-i)^2}{c^2}} \right) = 0$$

3B5
Ref
3A4
Ref
3B3

as with reference to 3A4 $\lim \varepsilon_x = \infty$, and with reference to 1B3 $\lim n = \infty$. Therefore the *blue shift* conflict is of infinite low value and the neutron process of the *Hydrogen* lasts for infinite time.

There are here two questions to be answered:
Can proton and electron processes be completed, while the neutron process is not? As electron process is the drive and each expanded drive continuous as neutron process, what the proportions of the already generated neutron processes in run are?
The answers are easy:
Proton and electron processes cannot happen without neutron process (even with no completion), otherwise the *Strong* and *Weak Interrelations* do not act and the element as such does not exist. Proton and electron processes are the expanding side, the neutron collapse as the balance-response is the other.
As the intensities of the proton and neutron processes clearly differ and the volume of the *Hydrogen* as element is growing and accumulating, this is only possible if the electron stage continues as neutron process.

The growing proportion of the *Hydrogen* element in the nature is coming from an elementary cycle with proton and electron processes but without the completion of the neutron process. *Hydrogen* as the last of the elementary cycles is generating, operating and/but accumulating for infinite time.
The communication of the *Hydrogen* with other elements means providing neutron process options for their electron process *blue shift* impact in $\lim n = \infty$ numbers and for infinite time.

The best demonstration of the communication of the *Hydrogen* is the case with the *Oxygen*:
-   whatever is the intensity of the *blue shift* drive and the generated neutron process portion of the *Oxygen*,
in communication with *Hydrogen*, the *Strong Interrelation* of the *Oxygen* process becomes fully balanced, as all its electron process surplus will drive available neutrons within the elementary structure of the *Hydrogen*.
The balanced status would result in solid elementary structure for the *Oxygen*, but the *blue shift* intensity surplus of the *Hydrogen* keeps the mix in liquid state – water.
All available electron process *blue shift* of the *Oxygen* do drive all available *Hydrogen* neutron processes, which will be covered by *Oxygen* protons. This way all earlier neutrons, part of the *Hydrogen* process become part of the *Oxygen* process. This way it would become solid, if there would not be electron process *blue shift* surplus within the composition. But it is, since the neutron process of the *Hydrogen* never ends.

The intensity of the electron process *blue shift* drive of the *Oxygen* is ahead of the intensity of the neutron process. This surplus is easily "swallowed" by *Hydrogen* neutron processes, to be covered by *Oxygen* protons. After the inflexion point the process is continuing as *Oxygen* process.

The process is escalating, since the formulated *Oxygen* protons continue as electron processes of increased intensity and surplus again, driving all available *Hydrogen* neutron processes again, turning them into *Oxygen* proton... and so on.

This means, the electron process of the *Oxygen* has been fully utilised, with no more surplus. The *Strong Interrelation* of the *Oxygen* corresponds to its standards, but the produced electron process *blue shift* is permanently taken by the neutron process of the *Hydrogen*.

The proton and electron processes of the *Hydrogen,* the existing constant *blue shift* surplus are not just obligatory conditions to the neutron process of the *Hydrogen*, but keep water in liquid status. Elementary relations of water are stabilised, with the existing specific features of the *blue shift* surplus of the *Hydrogen*.

Any external *blue shift* will have its impact to the *Quantum Membrane* of the water, kept in fact loaded by the electron *blue shift* surplus of the *Hydrogen*.

External impacts intensify the *Quantum Membrane* and the cycle and all processes go with increased quantum communication.

Fire is the best example:

Fire is electron process *blue shift* conflict of infinite intensity.

The electron process *blue shift* surplus of the *Oxygen* alone would feed this conflict. The *Hydrogen* within the water, with all its neutron process potential of infinite *low* intensity and of infinite numbers however takes all available *blue shift* surplus of the *Oxygen* away from the *blue shift* conflict of the fire. And the *fire* is off.

The *blue shift* impact of the *Hydrogen* is of infinite low intensity. This cannot feed a process needing an impact of infinite high *blue shift* intensity.

With the intensification of the water process, *fire* is not just losing its feeding *Oxygen* component, which is incorporating as *blue shift* drive and proton cover into water, but the intensity loss reduces the *blue shift* conflict, the reason of fire as well.

Water is with constant *blue shift* surplus, but the generating *Quantum Membrane* is very sensitive, as the *blue shift* load corresponds to the infinite low value of the intensity of the neutron collapse.

If the *blue shift* surplus of the water is taken away by cooling, the elementary status becomes solid.

There is no change in the elementary communication and structure of the *Hydrogen*, just the impact of the existing *blue shift* surplus taken away by cooling.

The relation of the *Hydrogen* to *Carbon* element is similar to that of the *Oxygen*, just as *Carbon* is with electron process *blue shift* surplus of less intensity the relation has different forms at normal temperature, depending on the proportions of the components. *Hydro-Carbons* may exist in gaseous, liquid and solid states. With external heating solid *Hydro-Carbon* become liquid, and liquid become gaseous.

The relation of the *Hydrogen* to *Carbon* minerals is different, as *Carbon* minerals are of solid status. The *blue shift* surplus of the *Hydrogen* communicates with the elementary structure of the minerals and because of this *blue shift* surplus, minerals become liquid and called *oil*.

3.1
The drive of the process in *Hydrogen-Carbons* is the *blue shift* surplus of the *Hydrogen*.

The elementary structure of *Carbon* minerals are close to equilibrium but with electron process *blue shift* deficit. This way they are in solid status. The *blue shift* drive of the *Hydrogen* communicates with the *Carbon* minerals. This communication may result in various forms even at normal temperature, depending on the proportions of the components. *Hydro-Carbons* may exist in gaseous, liquid and solid states.

What is the reason *Hydro-Carbons* feeds fire?
The elementary structure of the *Carbon* element is with quasi *blue shift* surplus (contrary to *Carbon minerals*), but close to proton-neutron process equilibrium state. Therefore *Carbon* is sensitive to quantum communication and in fact *pure Carbon* element as such does not exist in the nature just in minerals, in composition with other elements.
Coal minerals utilise all available electron process *blue shift* surplus of *Carbon* element. Therefore coal minerals are of solid structure, with integrated electron process *blue shift* intensity deficit.

The *blue shift* impact of the *Hydrogen* communicates with *Carbon* minerals and drives their neutron processes. This results in generation of electron process *blue shift* surplus in *Carbon* minerals free and available for quantum conflicts. Electron process *blue shift* impact with proton process cover, free in *Carbon* minerals drives *Hydrogen* neutrons of infinite low intensity, resulting in *Carbon* elementary structure after the inflexion point.
The dynamic balance is establishing here as well, just the starting point is the developing *blue shift* surplus of *Carbon* minerals. (While in the case of water this happens on the basics of the *blue shift* demand of the *Hydrogen*.)

Fire is *blue shift* conflict of infinite intensity and its impact on *Hydro-Carbons* is twofold:

(1) The *blue shift* conflict of infinite high intensity destroys the elementary structure of the *Carbon* minerals, established on the basics of the *Carbon* element.

   As *Carbon* and *Hydrogen* are with *blue shift* surplus, the fire as *blue shift* conflict becomes being much more conflicting.

   The destruction of *Hydro-Carbons* might need time, depending on their original state.

(2) The *Hydrogen*, released by the *Hydro-Carbon* conflict is communicating with the *Oxygen*, the main feeding component of the fire process as well. This communication results in generation of water within the fire.

   There are here again couple of options, depending on the size of the fire conflict. The generating water either subject to immediate vaporisation or flows away. There is no chance in the practice for extinguishing fire because of the generated water.

Pure *Carbon* element is the one closest to the elementary equilibrium state but still with slight electron *blue shift* surplus.

The diamond status is an easy one. The *blue shift* surplus of the *Carbon* in the structure of a diamond is in quasi balance with the *blue shift* deficit of the other elements of the mineral structure: It is very difficult to impact a $\varepsilon \cong 1$ balance of the integrated proton-neutron processes of the diamond.

With all other *Carbon*-minerals ε_C the intensity coefficient of the electron process is different and more than 1. Even anthracite, the "highest" quality coal is with *blue shift* deficit. Graphite is composition of *Carbon* with minerals, resulting in significantly reduced common *blue shift* effect. This massive *blue shift* deficit grants unique characteristics to *graphite*: taking electron process *blue shift* (good conductivity), being not sensitive to *blue shift* conflicts (thermal resistance). Graphite is naturally focused for gathering *blue shift* to improve its existing *blue shift* deficit.

Carbon-minerals with *blue shift* deficit need the "support" of the *blue shift* surplus and proton cover of the *Hydrogen* and will compose together liquid or gaseous statuses.

The intensity of the electron process of *Carbon*-minerals is:

3C2
$$\frac{dmc^2}{dt_i \varepsilon_i^m \varepsilon_x^m}\left(1 - \sqrt{1 - \frac{(c-i)^2}{c^2}}\right) = \frac{dn^m}{dt_i \varepsilon_i^m \varepsilon_x^m}q$$

The intensity of the electron process of the *Hydrogen* is:

3C3
$$\frac{dmc^2}{dt_i \varepsilon_i^H \varepsilon_x^H}\left(1 - \sqrt{1 - \frac{(c-i)^2}{c^2}}\right) = \frac{dn^H}{dt_i \varepsilon_i^H \varepsilon_x^H}q$$

The intensity of the electron process *blue shift* and the proton process surplus of the *Hydrogen* will be used by the neutron process of *Carbon*-minerals. The *blue shift* surplus available in the *Hydrogen* will be reduced.

With the use of the surplus of the *Hydrogen* electron process *blue shift*, the intensity of the electron process of the *Carbon*-mineral, the *Hydro-Carbon* will be also with *blue shift* surplus:

$$\frac{dmc^2}{dt_i \varepsilon_i^X \varepsilon_x^X}\left(1-\sqrt{1-\frac{(c-i)^2}{c^2}}\right) = \left(\frac{dn^m}{dt_i \varepsilon_i^m \varepsilon_x^m}+\frac{dn^H}{dt_i \varepsilon_i^H \varepsilon_x^H}\right)q \qquad 3C4$$

the *blue shift* is increased, since obviously: $\dfrac{dm}{dt_i \varepsilon_i^X \varepsilon_x^X} > \dfrac{dm}{dt_i \varepsilon_i^m \varepsilon_x^m}$ 3C5

Comments to the equations above are:
  (1) Formulas with intensity values written in are for the expression of the full balance. For characterisation of the event, intensity values have to be left out, otherwise all processes would be identical;
  (2) The time frame of all electron processes are the same: dt_i at speed level of $i = \lim a\Delta t = c$. The intensity increase is equivalent to more intensive *Quantum Membrane* electron process (*blue shift*) impact accelerating the collapse of the neutron process of *Carbon*-minerals.

There are here two points to be noted:

Diamonds with their full balanced status will never "communicate with *electrons*"; they will never be in any *blue shift* conflict. They are not flammable. [This is the reason diamonds are so good heat conductors and at the same time electrical isolators.]

Contrary to this, *graphite* is with *blue shift* deficit.

Graphite is *Carbon*-mineral with deep *blue shift* deficit. This makes them not flammable in normal circumstances.

3.2
The *blue shift* surplus of the *Oxygen* is the drive in *water*

S.
3.2

Oxygen has the strongest electron process *blue shift* impact of the Periodic Table driving the neutron process of the *Hydrogen*.

Hydrogen is with electron process *blue shift* surplus. *Hydrogen* has a quasi open cycle, with never ending neutron process of *infinite low* intensity.

In *Oxygen-Hydrogen* communication the overwhelming electron process *blue shift* surplus of the *Oxygen*, reason of its gaseous status disappears. It is utilised by driving the neutron processes of the *Hydrogen*, which after the inflexion point results in *Oxygen* elementary processes. The electron process *blue shift* intensity surplus of the new born *Oxygen* will be utilised the same way. This way the *Oxygen* process in mix with *Hydrogen* is speeding up, fully used and escalating.

The *Strong Interrelation* of the *Hydrogen* cannot remain without neutron process. For the driven neutron processes, as response and elementary communication, *Hydrogen* is taking neutron processes from the *Oxygen*, to be driven by its electron *blue shift* surplus and to be covered by its proton process.

The electron process drive and the proton process cover are the ones determining how the process will be continuing (what element will be re-burning) after the inflexion point.
The number and the volume of the *Hydrogen* process this way remains unchanged, while the *Oxygen* process in fact is speeding up.

The more of the neutron processes of the *Hydrogen* has been driven, the more are taken from the *Oxygen* process as well. *Oxygen* will use its all available *blue shift* surplus for driving *Hydrogen* neutrons. The liquid status is guaranteed by the *blue shift* surplus of the *Hydrogen* driving neutron processes of the *Oxygen*.

The mass-energy balance has not been destroyed.
Proton processes of the *Oxygen* will cover *Hydrogen* neutron processes. At the same time the electron process *blue shift* surplus of the *Hydrogen* will always drive neutrons taken from the *Oxygen*. This way the cover capacity of the proton process of the *Hydrogen* can also be utilised.

At one point the dynamic balance of the two processes is established and the water is of full energy. The consequences of the communication of these two elements are of significant importance:

> *One of the consequences* is the permanent utilisation – by the neutron process of the *Hydrogen* – of the electron process *blue shift* surplus and proton process cover capacity of the *Oxygen*; and this way the permanent rebirth of *Oxygen*.
> The neutron process of the *Hydrogen* can utilise *Oxygen* process *blue shift* surplus and proton process cover of infinite capacity in normal *Earth* circumstances.
> Water in fact is rebirth of the *Oxygen* in liquid format, full of utilised and re-established *Oxygen*.
> The liquid status is ensured by the constant *blue shift* surplus of the *Hydrogen*. The more is the electron process *blue shift* drive the *Oxygen* provides, the more is the water.
> The *other consequence* of the *Oxygen-Hydrogen* communication is that the electron process *blue shift* demand of the *Hydrogen* cannot be fully met.

In a certain case, if water is with electron process *blue shift* surplus more than that can be utilised by the *Hydrogen*, water falls apart into *Oxygen* and *Hydrogen*. This unnatural process is happening when breaking water by the *blue shift* impact of electricity.

3.3
Reaction of *water, Hydrogen* and other elements

Water is with *blue shift* surplus, generated or developed by the *Hydrogen.* Why the *Hydrogen?*
Because the *Oxygen* cycle of the water is speeded up: All available electron process *blue shift* surplus of the *Oxygen* is driving *Hydrogen* neutrons, while the number of the neutron processes of the *Hydrogen* in fact is unchanged.

In communication with other elements, the electron process *blue shift* surplus of the water generates *blue shift* conflict and loads the *Quantum Membrane.*

In the case of elements with *neutron process dominance* (like *Mg, Al, P, Na, Cl, Cu, Fe* and many others) the communication with water is limited or missing.
These elements are with electron process *blue shift* deficit, and therefore cannot drive the *Hydrogen* neutrons of the water. At the same time the low intensity of the electron process of the *Hydrogen* cannot "compete" with the higher internal intensity of the electron processes of these elements. There is no *blue shift* conflict is developing, a drive which would contribute to dissolve these elements.
The communication is limited and is only about the *blue shift* surplus of the *Oxygen* content of the water: the case is *Oxidation* of the surface.

If elements are with proton process dominance, the communication is of different nature. Apart of the *Oxygen,* there are only a couple of elements with proton process dominance: *Ca, C, Si, S, Ni, He.*
In a mixture of water and these elements the *Hydrogen* and also the *Oxygen* are participating in the communication.

The listed elements are to be found in the nature in minerals – as streaming to be balanced – in composition with other less *blue shift* intensive elements. The quasi solid, powder status of these minerals, still with integrated electron process *blue shift* deficit, can be dissolved and made liquid by the *blue shift* surplus of the *Hydrogen* content of the water.

Water, dissolving these minerals by the *blue shift* surplus of the *Hydrogen* also uses the elementary process of the *Oxygen* with all its quantum communication capabilities. If this *blue shift* impact is enough in its volume and intensity to mobilise the utilised *blue shift* impact of the element within the mineral, the consequence is *blue shift* conflict and increased temperature of the mix-liquid.
The *Quantum Membrane* of the liquid becomes of increased *blue shift* conflict and load. The communication between the *Oxygen,* the *Hydrogen* and the elements starts.

Not just the liquid but also the quantum space, the *Quantum Membrane* around the liquid becomes impacted by the *blue shift* (energy) potential of the mix. It can be measured. The *blue shift* impact is provided as that happens in the case of ice: The external environment is taking off the *blue shift* as a kind of cooling process of the mix.

If the liquid solution now is mixed with minerals with neutron dominance, close to equilibrium state, the mixture becomes neutral but with certain built in electron process *blue shift* load.

Elements with electron process *blue shift* surplus within the mix will be communicating and the *Hydrogen*, built in, provides the *blue shift* load. The internal process can be followed, but the mix externally will be neutral.

S.
3.4

3.4
Acid – base relations

The internal relation of the *Hydrogen* and the *Oxygen* makes elementary compositions of minerals *bases* and *acids* as they are.
Both *acids* and *bases* are *Hydrogen* dependent, but in *bases Oxygen* is directly connected to *Hydrogen*.

The intensity of the electron process *blue shift* of the *Hydrogen* process is infinite low and the *Hydrogen* process never ends. The *Oxygen* process is speeding up, but the intensity of the *Hydrogen* process remains unchanged.

3D1
$$\lim \varepsilon_H = \frac{\varepsilon_p}{\varepsilon_n}\sqrt{1 - \frac{(c-i)^2}{c^2}} = 0 \, ; \qquad \varepsilon_O = \frac{\varepsilon_p}{\varepsilon_n}\sqrt{1 - \frac{(c-i)^2}{c^2}} = x > 1$$

If the starting volumes of *Hydrogen* and *Oxygen* are both *A,* after a certain time period, while the *Hydrogen* content remains still *A,* the *Oxygen* becomes: *xA*. In each *Oxygen* cycle the intensity of the proton process is increasing, generating more electron processes and *blue shift* drives for the neutrons of the *Hydrogen* to be driven the same way!

The difference in *bases* and *acids* are
➤ *Bases* are with electron process *blue shift* <u>demand</u>.
-   The *blue shift need* of the *Hydrogen* in *bases* is *softened* through its (OH) content by *Oxygen*.
    Blue shift is needed for driving the neutron process of the *Hydrogen*. The neutron process of the *Hydrogen* in water is fully utilising the *blue shift* surplus of the *Oxygen*. Here in bases the electron process *blue shift* surplus of the *Oxygen* is only partially driving it.
➤ *Acids* have increased electron process *blue shift* <u>surplus</u>.
-   There are two factors here: (1) the neutron process of the *Hydrogen* has not been fully driven by *Oxygen* electron process *blue shift* plus (2) the *blue shift* surplus of the *Hydrogen* is still available for use.

There are variants for establishing acids:

- the *blue shift* of the increased *Oxygen* content in some acids is either in communication with other elements (like H_3PO_4; H_2CrO_4; H_3BO_3);
- or acting in line with the elements with *blue shift* surplus (like H_2SO_4; HNO_3; HSO_3F; CH_3SO_3H; $C_6H_8O_7$ and others);
- or the mix is acting without *Oxygen* content at all, leaving the *Hydrogen* without efficient cover (like $HSbF_6$; HBF_4; HPF_6; and others).

The definitions for *bases* and *acids* are relativistic, as both in fact are in *blue shift* demand and at the same time in surplus. In the case of *blue shift* demand, the surplus is softened, in the case of surplus, the *blue shift* demand strengthened.

Acids and *bases* can optimise their "strengths" and "weaknesses" and utilise the missing *blue shift* surplus available.

Processes called in the chemistry as *protonation* and *de-protonation* are about adding and removing *Hydrogen* proton (as called in conventional science H^+ ion) to and from atoms, molecules and ions.

3.5

Fire

Blue shift conflict of infinite intensity destroys elementary structures.
What does this mean?
Elementary cycles have been speeded up:
The proton- electron- neutron- proton- electron- neutron- proton... infinite cycle is speeding up. Electron process *blue shift* conflict is generating. External electron process *blue shift* feeds the conflict, resulting in light and heat impacts.

Fire is *blue shift* conflict of infinite intensity: conflicting electron process *blue shift* interactions of elements or compounds with *blue shift* surplus.
But the *blue shift* surplus potential itself is not enough for having fire.
"Physical" conflict between the *blue shift* impacts of the electron processes is necessary. The symptoms of the conflict are sparking light effects and heat generation.

Balance problems between proton and neutron processes may also result in *blue shift* conflict. Elements in normal circumstances without electron process *blue shift* surplus and with neutron process dominance can also be subject to *blue shift* conflict this way. The best example is the conflict of the *Uranium* isotope in nuclear reactors.

The conflict destroys the element and impacts the *Quantum Membrane.*
The size of the conflict corresponds to the number of the conflicting electron process *blue shift* impacts:

3E1
$$X \frac{dmc^2}{dt_i \varepsilon} \left(1 - \sqrt{1 - \frac{(c-i)^2}{c^2}} \right) = X \frac{dn}{dt_i \varepsilon} q \quad \text{which can be written as}$$

3E2
$$\text{equal to} \quad = \frac{dmc^2}{\frac{1}{X} dt_i \varepsilon} \left(1 - \sqrt{1 - \frac{(c-i)^2}{c^2}} \right) = \frac{dn}{\frac{1}{X} dt_i \varepsilon} q$$

3E3   The coefficient in the denominator is equivalent
to an acceleration to certain speed value u $\frac{1}{X} = \sqrt{1 - \frac{u^2}{c^2}}$

The conflict at this level also results in the increase of the intensity of the neutron and proton processes.
The neutron process will be:

3E4
$$X \frac{dmc^2}{dt_i} \sqrt{1 - \frac{(c-i)^2}{c^2}} \left(1 - \frac{1}{\sqrt{1 - \frac{v^2}{c^2}}} \right) = \frac{dmc^2}{dt_i \sqrt{1 - \frac{u^2}{c^2}}} \sqrt{1 - \frac{(c-i)^2}{c^2}} \left(1 - \frac{1}{\sqrt{1 - \frac{v^2}{c^2}}} \right)$$

increased intensity of the neutron process

The *blue shift* conflict this way can be described as effect of acceleration up to a certain speed value. The higher is the conflict the more intensive is the supposed acceleration and the end speed.
Acceleration from external energy source in general means:

3E5
$$w = \frac{dmc^2}{dt} \left(1 - \frac{1}{\sqrt{1 - (u^2/c^2)}} \right)$$

The *blue shift* conflict as fictional speed increase intensifies the full cycle.
The acceleration not just intensifies the complete elementary processes but also changes the intensity characteristics of the elements.
$i = \lim a \Delta t = c$, the elementary time frame of the electron process of

3E6   elements cannot be changed but $dt_i \sqrt{1 - (u^2/c^2)}$ is equivalent to

the <u>slow-down</u> of the *time flow* with significant intensity increase!

Intensified *blue shift* conflict and slowed down time flow means the increase of c, the speed of local *quantum communication*.

Ref   As 3D2 proves, the frequency of the *Quantum Membrane* impact depends on
3D2   the size of the conflict – the number of the acting *blue shift* impacts:

3D7
$$\frac{dn}{\frac{1}{X} dt_i \varepsilon} q = \frac{dn}{dt_i \sqrt{1 - (u^2/c^2)} \cdot \varepsilon} q = f_x \cdot q$$

The conflict itself is from the increased *blue shift* impact.
The summarised electron process *blue shift* impact of all participants of the conflict well represents the case. The higher is the resulting conflicting *blue shift* value, the higher is the frequency of the quantum impact.

Once a conflict happens the conflicting relations usually disappear in the general practice: the reason, causing the conflict has been resolved by the conflict itself.

In the case of *fire* the conflict is permanent. The key here is the unique and electron *blue shift* surplus of the air, mainly the *Oxygen*.

Oxygen has its aggressive *blue shift* impact to the elementary process of other elements and this way it feeds the conflict and makes it permanent.

The characteristics of this permanent conflict – *fire* – however depend on the intensity of the original conflict as well. In certain conflicting cases, like nuclear fission the intensity of the conflict is so high that there is no chance to any *Oxygen* impact to keep it permanent.

Feeding the conflict means keeping the intensity of the conflict (the event) high. One point is the acceleration itself and the other is to maintain a certain speed. The *blue shift* conflict accelerates the cycle but keeping it at this speed level also needs *blue shift*. It can be of different and even of less value. *Oxygen* provides this additional electron process *blue shift* source available for maintaining this increased intensity (fictional speed) value.

Fire means all electron, neutron and proton processes go with increased intensity. This is the reason of the experienced heat impact of fire: the increased intensity means more intensive quantum impact. The more intensive is the original intensity increase (at the spark of the conflict), the more intensive is the heat impact.

The intensity increase of the proton and neutron processes means online increase. For the electron process it means increased c, *speed of quantum communication,* as $i = \lim a\Delta t = c$ is of higher speed. The intensity increase for all processes means increased mass-energy exchange, as dmc^2 includes the speed of quantum communication for all components.

There is another condition for keeping the increased speed value of quantum communication of elementary process – *fire* – acting:
> the availability of the natural *blue shift* surplus of the element.

If the element is without the necessary *blue shift* capacity, the provided to the *blue shift* conflict by *Oxygen blue shift* "support" and proton process cover will be acting, but "swallowed" by the neutron collapse of the element itself and the intensity increase would be lost. Elements with *blue shift* deficit therefore can only be kept in *fire* if external heating is provided otherwise they cannot be made flammable at all in normal circumstances.

If the *blue shift* surplus is intensified, the conflict to the *Quantum Membrane* is increasing. This is the benefit of the incineration of elements: generating heat as result of the increased *blue shift* conflict to the *Quantum Membrane*.

The generation or the feeling of the heat however is not the conflict itself – that is the impact to the *Quantum Membrane*.

Whatever is the value of the initial *blue shift* conflict, *Oxygen* can only maintain a certain "speed" value equivalent (as above). The level of the heat generation therefore depends on the *blue shift* surplus of the element or the mix of the elements in *fire*.

The *Hydrogen* content of *water* in normal circumstances utilises all *blue shift* and proton cover available. Therefore water is excellent for *extinguishing fire*.

Fire as intensified (slowed down) elementary process also modifies the structure of the elementary composition. The acting electron process *blue shift* surplus of the *Oxygen* destroys the *blue shift* balance within the compounds (minerals) and provides the needed electron *blue shift* to the elements. Flue gases are molecules with *Oxygen*.
[*Hydro-Carbons* burn into CO_2 flue gases and water. Burning coal minerals may result in all *Oxides* of the elementary structure of the coal mineral.]

In the case of minerals with many elementary components, elements closer to the proton-neutron balance get accelerated easier and burn out into flue gases as result of the *Oxygen* interaction. Elements with heavy neutron processes need either higher external energy (additional *blue shift* intake) for making them capable to be in conflict and accelerated or will remain without acceleration as end products of the fire.

In the case of elements with damaged proton-neutron process balance – as *isotopes* are – fire deepens the *Quantum Membrane* impact of the isotope, independently of the type of the radiation of the isotope, *alpha*, *beta*, *gamma* or *X-ray*. The impact will be intensified as happens in equivalent slowed down time flow.

Therefore the radiation impact of isotopes increases in fire. At the same time, as the process is more intensive the half-life of isotopes is shortened. The flue gases in this case in composition with *Oxygen* become radioactive as well.

4
Neutron, alpha, beta, gamma **and** *X-ray* **radiation and rehabilitation**   S.4

The mass-energy balance of elements is:

$$\left|\frac{dmc^2}{dt_p\varepsilon_p}\left(1-\sqrt{1-\frac{i^2}{c^2}}\right)\right|=\left|\frac{dmc^2}{dt_n\varepsilon_n}\xi\sqrt{1-\frac{(c-i)^2}{c^2}}\left(\sqrt{1-\frac{i^2}{c^2}}-1\right)\right|$$ 4A1

The formula represents the end stages of the proton and neutron processes, at v=i.

The balance is only valid in its complex format, as above, with *time* and *intensity* parameters combined and with *entropy* indicated.
Otherwise (without intensity values) the equality is usually not valid:

$$\frac{dmc^2}{dt_p}\left(1-\sqrt{1-\frac{i^2}{c^2}}\right)\neq\frac{dmc^2}{dt_n}\sqrt{1-\frac{(c-i)^2}{c^2}}\left(\sqrt{1-\frac{i^2}{c^2}}-1\right)$$ 4A2

The intensities of the proton and neutron processes are not equal. They are not equal even in the most balanced case, but they are interdependent. This interdependence is the key for having elements with different values of *blue shift* surplus and deficit and different elementary characteristics.

There are *two* important messages formulated in the above equations:
➢ The balance of the transformation of mass into energy (proton process) and the re-transformation of the energy into mass (neutron process) - the *Strong Interrelation* between the proton and neutron processes is the basic cohesion force of the elementary structure.
➢ The electron process, the *Weak Interrelation,* is the drive of the process.
The harmony of these *two* interrelations gives the intensity formula and establishes, in fact, the element:

$$\left(\frac{dm}{dm}\right)\frac{dt_n}{dt_p}=\frac{\varepsilon_p}{\varepsilon_n}\sqrt{1-\frac{(c-i)^2}{c^2}}=\varepsilon_e \quad\text{and}\quad \frac{\dot{m}_n}{\dot{m}_p}=\frac{dm}{dt_n}\frac{dt_p}{dm}=Z\,;\quad\text{and}\;\varepsilon_x=\frac{1}{Z}$$ 4A3

Each element has its certain *Z* (*event concentration*) value. Each element has its own electron process intensity coefficient (ε_x), characteristic to the mass-energy balance of the element. Deviations in these relations result in damaged balance – reason of *isotopes*.

With reference to 4A3, *dt* on the left-hand side (with *p* and *n* prefixes) characterises the intensity of the proton and neutron processes.
ε and *1/Z* characterise the element. ε and *1/Z* are without dimension.
Once the equalities in 4A3 are disrupted, the element is damaged.

The intensity of the electron process (reciprocal to the value of the intensity coefficient) characterises the mass change: higher electron process intensity means higher mass change drive for the unit period of time (within the constant time system of the electron process).

If the natural balance is broken, the symptoms are the following:

4B1   *1. Blue shift conflict:* $\dfrac{dt_n}{dt_p} > \dfrac{\varepsilon_p}{\varepsilon_n}\sqrt{1-\dfrac{(c-i)^2}{c^2}}$ meaning $Z < Z_{normal}$

4B1 means the intensity of the proton process is more than that is for the standard (normal) element should be. The less intensity of the neutron process would be in principle the other option, but the neutron process is always driven by electron process, and so it is dependent.

But the *Strong Interrelation* – the balance between the proton and neutron processes – is still controlling the case. Meaning: no disruption in the continuity of the reciprocal relation (proton coverage to neutron collapse) of the two processes.

The deviation, however, is not without consequence: the electron process surplus, result of the increased intensity of the proton process creates increased *blue shift* conflict within the elementary communication.

The intensity of the electron process contrary to the developing surplus and conflict however will be of less value.

The symptoms of the electron process *blue shift* surplus are:

(1) developing *heat* impact and (2) less neutron process mass impact, as direct consequence of the decreased intensity of the electron process.

Up to certain limits *nature* (the *Strong Interrelation* of the element) tolerates the imbalance.

Typical examples are isotopes of $^{92}_{235}U$, $^{94}_{239}Pu$ and $^{94}_{241}Pu$.

The less intensive neutron process results in less intensive collapse, mass impact (weight):

4B2 $$\frac{dmc^2}{dt_n^{U235}}\sqrt{1-\frac{(c-i)^2}{c^2}}\left(\sqrt{1-\frac{i^2}{c^2}}-1\right) < \frac{dmc^2}{dt_n^{U238}}\sqrt{1-\frac{(c-i)^2}{c^2}}\left(\sqrt{1-\frac{i^2}{c^2}}-1\right)$$

For good measure, it shall be noted that the full weight of elements is the integrated impact of the neutron and proton processes at the inflexion point:

4B3 $$G = \left|\frac{dm}{dt_p}\left(1-\sqrt{1-\frac{i^2}{c^2}}\right)\right| + \left|\frac{dm}{dt_n}\sqrt{1-\frac{(c-i)^2}{c^2}}\left(\sqrt{1-\frac{i^2}{c^2}}-1\right)\right|$$

In the case of $^{92}_{235}U$, the limit is up to 0.7% concentration. The electron process *blue shift* surplus of the $^{92}_{235}U$ isotope creates heat, but it has been taken away by the surrounding environment.

(Elements vary. There are ones with natural electron process *blue shift* surplus, others with deficit or quasi balance. The imbalance shall be understood as deviation relative to the standard status of elements.)

Alpha radiation: $\dfrac{dt_n}{dt_p} >> \dfrac{\varepsilon_p}{\varepsilon_n}\sqrt{1-\dfrac{(c-i)^2}{c^2}}$ meaning $Z << Z_{normal}$ 4C1

The deviation between the intensities is more significant.
The intensity of the proton process and the electron process *blue shift* conflict are further increased. It threatens the *Strong Interrelation* and with that the integrity of the element.

In order to manage the intensity difference and to try to resolve the imbalance, element releases complete mass-energy transformation cycle with natural *blue shift* surplus: The release of a complete cycle with increased mass-energy transformation (proton process) and deceased energy-mass re-transformation (neutron process with *blue shift* surplus but of less intensity) obviously improves the imbalance, since part of the non-balance is released.
The smallest complete cycle is that of the *Helium* element.
Hydrogen with its infinite long and infinite low intensity neutron process is a "never" completing cycle and cannot be used as release and solution.
The release of *Helium* is *alpha* (α) radiation.

Alpha (α) radiation is release of the complete mass-energy transformation cycle of the *Helium*. It means released ("flying" away) complete processes with kinetic energy. It has a short range effect, but because of the high mass impact, it might be very damaging.

3. *Beta* radiation: $\dfrac{dt_n}{dt_p} \neq \dfrac{\varepsilon_p}{\varepsilon_n}\sqrt{1-\dfrac{(c-i)^2}{c^2}}$ meaning $Z \neq Z_{normal}$ 4D1
4D2

The balance of the proton-electron-neutron processes has been destroyed. The non-balance may happen in various ways, but the main point is that the mass-energy balance of the cycle is destroyed.

either $\dfrac{dm}{dt_n^{isotope}} < (or) > \dfrac{dm}{dt_n}$ or $\dfrac{dm}{dt_p^{isotope}} > (or) < \dfrac{dm}{dt_p}$ 4D3

This is damage in the *Strong Interrelation*, within the balance of the mass-energy-mass… endless transformation.

$$\frac{dmc^2}{dt_p\varepsilon_p}\left(1-\sqrt{1-\frac{i^2}{c^2}}\right) \neq \left|\frac{dmc^2}{dt_n\varepsilon_n}\sqrt{1-\frac{(c-i)^2}{c^2}}\left(\sqrt{1-\frac{i^2}{c^2}}-1\right)\right| \qquad 4D4$$

If the proton $\dfrac{dmc^2}{dt_p}$ and the neutron process $\dfrac{dmc^2}{dt_n}\sqrt{1-\dfrac{(c-i)^2}{c^2}}$
process intensity   intensity at the
at the start is   inflexion point is

the difference between the intensities of the
proton and neutron processes in normal cases $\left|\dfrac{dmc^2}{dt_i\varepsilon_x}\left(1-\sqrt{1-\dfrac{(c-i)^2}{c^2}}\right)\right|$
corresponds to and can be expressed by the   4D5
electron process, which is now broken

Fermi formula describes the proton-neutron-electron *evolution* (without calculating with the developing energy quantum and quantum entropy.)

4D6   The *Fermi* decay is: $n \rightarrow p + |e| + |\upsilon|$,

4D7   which is equivalent to $n \rightarrow p + |\beta| + |\upsilon|$

In the case of the disruption of the balance, value of β obviously depends on the polarity, on the positive (*positron*) or the negative (*electron*) signs of the deviation of the decay (balance).

The *neutrino* or *antineutrino* parts are possible supplementary corrections to the balance.

As corrective action, an *isotope* with this kind of damaged balance either provides or takes *blue shift* – in conventional terms: either *releases* or *takes* electron process effect(s) from others.

At the inflexion point: $\dfrac{dm}{dt_p} = \dfrac{dm}{dt_n}$;   since $dt_o = dt_p = dt_n$

and the intensity difference in general is addressed by $\varepsilon_x = \dfrac{\varepsilon_p}{\varepsilon_n} \sqrt{1 - \dfrac{(c-i)^2}{c^2}}$

ε_x - is the intensity coefficient of the electron process, *at standard dt_i* time system of $i = \lim a\Delta t = c$ speed *and standard intensity* of ε_i

(the higher is ε_x the less is the intensity of the electron process).

The intensity of the electron process is combined by two components:

ε_i the intensity, characterising the time system of the electron process;

ε_x the intensity coefficient, characterising the relation of the proton and the neutron process intensities of the element.

The absolute formula of the electron process is:

4D8
$$W_{electron} = \frac{dmc^2}{dt_i \varepsilon_i \varepsilon_x}\left(1 - \sqrt{1 - \frac{(c-i)^2}{c^2}}\right)$$

With reference to the corrective actions of elements:

Releasing electron process *blue shift* impact (causing so-called β^+ *positron* radiation) – improves the consequences of the increased intensity of the proton process, when electron process *blue shift* surplus and conflict is generating while the intensity of the electron process is of less intensity (energy) than that is necessary for the normal standard proton-neutron relation of the element. This radiation variant case could be positioned between the *blue shift* conflict and *alpha* radiation.

This obviously is an impact to the neutron process as well, since the neutron drive with the release will be less. The gradient of the proton process intensity will be less and the generation of the electron process intensity will also be less.

Taking electron process *blue shift impact and proton process cover from other elements* – improves the missing elementary balance of an isotope. This is the case when the proton process of the isotope is of less intensity and cannot generate electrons in necessary numbers to collapse and cover the neutron process of the element. (The neutron collapse of the standard element with stable structure needs more electron process intensity drive and proton cover than that is available within the isotope).

The isotope with missing electron process *blue shift* drive and proton process cover would not bring the neutron process to the inflexion point. Neutron process would turn around and the collapse would turn into expansion (taking over proton function) for compensating the damage – as it happens in the case of *gamma* radiation.

Taking electron process *blue shift* drive as electron or β^- *radiation* and proton process cover from other elements completes the neutron collapse. The intensity of the inflexion point in this case however generates proton process of that certain element, providing the *blue shift* drive and the proton process cover. The rehabilitation of the isotope this way is causing damage in the elementary structure of the "donor" element. The elementary damage has not disappeared, just transformed to another element.

With reference to 4A1:

$$\left| \frac{dmc^2}{dt_p \varepsilon_p} \left(1 - \sqrt{1 - \frac{i^2}{c^2}} \right) \right| = \left| \frac{dmc^2}{dt_n \varepsilon_n} \sqrt{1 - \frac{(c-i)^2}{c^2}} \left(\sqrt{1 - \frac{i^2}{c^2}} - 1 \right) \right|$$

Ref 4A1

$\dfrac{dm}{dt_n} \neq \dfrac{dm}{dt_p}$ The intensities of the neutron collapse and the proton expansion of the element are different. The standard value of the relation is the *characteristic* of the element.

The full rehabilitation of the isotope without external β^- support needs time, depending on the intensity of the transition of the neutron-proton processes of the element at the inflexion point.

There are no flying mass particles – electrons (or positrons) and protons released or receipt by isotopes – in *beta* radiation. Releasing (positron) or taking electron processes means impacting the *Quantum Membrane* by the intensity of the electron process. *Quantum Membrane* transfers the impact and *beta* radiation either provides electron process *blue shift* impact to other elements (β^+) or takes electron process *blue shift* impact from other elements (β^-). Proton cover is provided by the *Strong Interrelation* in line with the continuity rule of the *Quantum System of Reference.*

The single impact is: $\dfrac{dmc^2}{dt_i \varepsilon_x} \left(1 - \sqrt{1 - \dfrac{(c-i)^2}{c^2}} \right)$ = electron *blue shift*

Ref. S.8.4

With reference to 4D8, the time system of the electron process is constant, but the difference in intensities, results in different (increased or decreased) frequencies of the electron process.

Intensity can be demonstrated as certain increased or decreased mass impact:

4D9
$$beta = \frac{dmc^2}{dt_i \varepsilon_x}\left(1 - \sqrt{1 - \frac{(c-i)^2}{c^2}}\right)$$

Isotopes are natural components of the infinite chain of elementary cycles. Going through endless mass-energy-mass transformation, elements produce *beta* isotopes. Standard proton-neutron process intensity relations are forming as the result of *natural* progress, evolution, change.

The significance of this type of radiation is that this is not just about the electron process rather about the proton and neutron processes as well.

The damaged proton/neutron balance means that
-   release of *positrons* also means proton process intensity surplus of the element (independently whether there is any need);
-   *electron* process *blue shift* demand (of the neutron process) also means proton process cover need for the completeness of the elementary cycle.

Fig. 4.1 demonstrates the natural richness in isotopes. As example, *Carbon* has 15, *Caesium* has 39 isotopes.

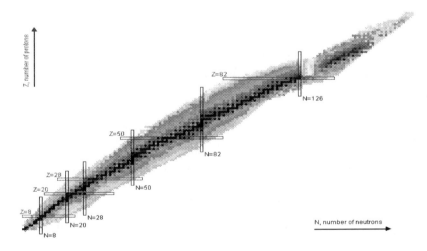

Fig.
4.1   Fig.4.1

The meaning of Z in the figure is different than the meaning used in this book. Z in this book is the relation of proton and neutron process intensities.

Elements with standard and balanced consolidated proton-neutron process intensity relation status are the ones on the straight line within the range of data, all with the characteristics of isotopes statuses.
(In the region toward the end with heavy elements the diagram is not precise, because of the missing or not available by measurement data. The diagram is taken from the Internet.)

Electron and proton processes are connected.

The impact of *beta* radiation is also equivalent

- either to *loss* on proton process intensity (in the case of β^-),

- or to possible proton process interference (in the case of β^+) as well.

For this reason the impact of the *electron* radiation (β^-) is more damaging.

The close reference of the electron process to the *Strong Interrelation* makes the effecting distance of the *beta* radiation so limited and short.

Beta rehabilitation also means limited in volume re-birth of those elements receiving *blue shift* drive and proton cover: in the case of β^- this is the element of the isotope; in the case of β^+ this is the external element of the impact from the isotope.

This way, as side effect of *beta* radiation neither β^- isotope in its rehabilitated format, nor other elements impacted by β^+ isotope remain "purely" the same. Rehabilitated β^- isotopes will be "infected" by elements providing *blue shift* drive and proton cover. β^+ isotopes, impacting other elements by their released *blue shift* surplus and proton cover will be in elementary communicating with the elements of the impact.

Infinite cycles of mass-energy-mass... transformations rehabilitate β isotopes of both formats without any "environmental" contribution.

4. <u>*Gamma* radiation</u>: $\dfrac{dt_n}{dt_p} << \dfrac{\varepsilon_p}{\varepsilon_n}\sqrt{1-\dfrac{(c-i)^2}{c^2}}$ meaning $Z >> Z_{normal}$ 4E1
4E2

Deviation from the natural elementary process, which is different than *alpha* or *beta* ways is the so-called *gamma* radiation.

This is the case when the electron process *blue shift* drive is so weak and the missing proton process intensity is so significant that the elementary process cannot provide the necessary proton process cover to the neutron collapse. The collapse of the neutron process is not completed.

The mass-energy transformation balance is seriously damaged.

Neutron collapse cannot remain without sufficient *Strong* proton mass expansion intensity cover *Interrelation*:

The proton process (1) does not generate the necessary electron process *blue shift* drive and (2) does not have sufficient intensity for covering the intensity demand of the neutron process. The element stays the same but the consequence of the proton process non-coverage is that the neutron process turns around and starts to expand as a proton would do.

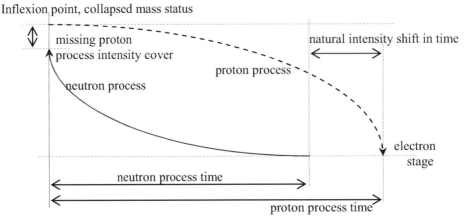

Fig.
4.2

Fig. 4.2

In absolute terms, it means that before the inflexion point, without full collapse, neutron turns into expansion.

The frequency of the impact of the unnatural *expanding acceleration* of the neutron depends on at what stage of the collapse the turnover happens. At the end stage of the neutron process, the *blue shift* of the neutron process against the *Quantum Membrane* is of high energy (frequency), since the "time count" of the process is slow: *gamma* (γ) radiation.

Gamma radiation is mass expansion against the *Quantum Membrane,* the result of the sphere symmetrical expanding acceleration of the neutron.
At the stage of the collapse of speed u and time system of $dt_{(i-u)}$, the generated *gamma "blue shift"* is a brutal impact to the *Quantum Membrane.*
The impact depends on the decrease of the speed and the time system of the start of the impact (the mass expansion).

4E3
$$w_\gamma = \frac{dmc^2}{dt_o\sqrt{1-\frac{(i-u)^2}{c^2}}}\sqrt{1-\frac{i^2}{c^2}}\sqrt{1-\frac{(c-i)^2}{c^2}}\left(1-\sqrt{1-\frac{[c-(i-u)]^2}{c^2}}\right)$$

There are no flying particles.
Gamma radiation is of high frequency *blue shift* impact to the *Quantum Membrane* generated by the damage of the *Strong Interrelation.*

Neutron does not have category *capacity*, since neutron process is driven by electron process *blue shift*. In the case of insufficient collapsing load from electron process *blue shift* – if the proton process does not generate sufficient number of electrons and does not provide the sufficient *"Strong"* intensity balance cover *"Interrelation"* of the collapse in full, the collapsed and accumulated neutron mass "capacity" will be expanding and with that will be generating *gamma* radiation.

<u>*X-ray* radiation:</u> $\dfrac{dt_n}{dt_p} < \dfrac{\varepsilon_p}{\varepsilon_n}\sqrt{1-\dfrac{(c-i)^2}{c^2}}$ meaning $Z > Z_{normal}$ 4F1
4F2

Bombarding certain elements, like *Fe, Ni, Co, Cu, Zr, Mo,* with electrons results in *X-ray* radiation.

Z, the event concentration of these elements is slightly above *1*. They are with natural neutron process intensity dominance. Bombarding them with electrons is additional massive *blue shift* impact to the surface regions of these metals. Provision of additional flow of electron process to their surface results in electron process *blue shift* conflict with the acting natural *blue shift* impact of the elements: electron process is in surplus, but the neutron process cannot be driven, since there is no proton process cover available.

The result is *X-ray* radiation = *blue shifted blue shift* with

$$\text{intensity:}\quad X-ray = \frac{dmc^2}{dt_i\varepsilon_e}\left(1-\sqrt{1-\frac{(c-i)^2}{c^2}}\right)^x \qquad \text{4F3}$$

x in the power addresses the multiplication of the *blue shift* impact.

X-ray radiation is an increased electron process *blue shift* impact to the *Quantum Membrane*. The frequency of the *blue shifted blue shift* impact depends on the electron process intensity of the bombarded elements, as this is the reflection point of the original external *blue shift* impact. The capacity of the impact depends on *x*, the exponent power of the equation.

4.1   S.
Neutron radiation   4.1

Within the scenarios of the elementary mass-energy imbalance, there is a case when the internal and natural compensation is impossible: the imbalance leads to the destruction of the element.

$\dfrac{dt_n}{dt_p} >>> ... \dfrac{\varepsilon_p}{\varepsilon_n}\sqrt{1-\dfrac{(c-i)^2}{c^2}}$ meaning $Z <<< ...Z_{normal}$ 4G1

When there is a certain concentration of the $^{92}_{235}U$ isotope (or $^{94}_{239}Pu$ or $^{94}_{241}Pu$ elements) within the natural $^{92}_{238}U$ element and/or under the effect of additional external *blue shift* impact (water, graphite), the *Strong Interrelation* cannot compensate for the imbalance and cannot keep the integrity of the element. $^{92}_{235}U$ is with electron process *blue shift* surplus.

The symptom of the conflict is the increased temperature of $^{92}_{235}U$.

The result of the increased *blue shift* conflict is that $^{92}_{235}U$ (or $^{94}_{239}Pu$ or $^{94}_{241}Pu$) falls apart into new elements with less atomic weight.

The birth of the new elements needs less *blue shift* drive intensity, since the natural intensities of the neutron processes of the new elements relative to the unit intensity of the proton process (in the middle range of the periodic table) are less than that was for $_{235}^{92}U$:

4G2
$$\frac{dm}{dt_n^U} > \frac{dm}{dt_n^X} \qquad \text{(The destruction of the \textit{Pu} isotopes results in the same effect.)}$$

With the growth of the periodic number – neutron intensities of elements are increasing.

4G3
$$\frac{dmc^2}{dt_p^U \varepsilon_p^U}\left(1 - \sqrt{1 - \frac{i^2}{c^2}}\right) = \frac{dmc^2}{dt_n^U \varepsilon_n^U}\,\xi_U\sqrt{1 - \frac{(c-i)^2}{c^2}}\left(\sqrt{1 - \frac{i^2}{c^2}} - 1\right)$$

4G4 below shows the electron process *blue shift* reserve of the neutron process of the $_{235}^{92}U$ balance for one of the resulting elements:

4G4
$$\frac{dmc^2}{dt_p^U \varepsilon_p^U} = \frac{dmc^2}{dt_n^X \varepsilon_n^X}\sqrt{1 - \frac{(c-i)^2}{c^2}} + \left(\frac{dmc^2}{dt_n^U \varepsilon_n^U} - \frac{dmc^2}{dt_n^X \varepsilon_n^X}\right)\sqrt{1 - \frac{(c-i)^2}{c^2}}$$

the *blue shift* intensity benefit of the less intensive neutron process of element X can be used:

4G5
$$e_i = \left(\frac{dmc^2}{dt_i^U} - \frac{dmc^2}{dt_i^X}\right)\left(1 - \sqrt{1 - \frac{(c-i)^2}{c^2}}\right)$$

[With reference to Section 9, quantum speed values in all formulas here are taken for equal. With different speed values, the *blue shift* reserve is even more increased.]

The source of "energy generation" is de facto the case that the new-born elements need less intensive electron process *blue shift* impact, as drive for their neutron processes. The available electron process *blue shift* intensity surplus and conflict is the basis of the energy generation.

The proton-neutron process intensity balance of the new elements of the fission, however, is heavily impacted: the newly created elements are with proton processes of higher intensity than the original element of the fission.

4G6
The "demand" in proton process intensity of the new elements is higher than it was for the $_{235}^{92}U$: $\qquad \dfrac{dm}{dt_p^U} < \dfrac{dm}{dt_p^X}$

Alongside the *blue shift* intensity surplus, result of the fission, as drive, the continuity of the proton-neutron mass balance, the *Strong Interrelation* must be kept: the newly born elements need additional proton mass expansion intensity cover and electron process *blue shift* drive from anywhere around to establish the elementary balance and complete the fission.

The proton and neutron process relations as per in 4G4, is:

4G7
$$\frac{dm}{dt_p^X \varepsilon_p^X} = \frac{dm}{dt_n^X \varepsilon_n^X}\,\xi_x\sqrt{1 - \frac{(c-i)^2}{c^2}}$$

The proton process need of the neutron process is:

$$\frac{dmc^2}{dt_p^X \varepsilon_p^X}\left(1-\sqrt{1-\frac{i^2}{c^2}}\right) = \frac{dmc^2}{dt_p^U \varepsilon_p^U}\left(1-\sqrt{1-\frac{i^2}{c^2}}\right) + \left(\frac{dmc^2}{dt_p^X \varepsilon_p^X}-\frac{dmc^2}{dt_p^U \varepsilon_p^U}\right)\left(1-\sqrt{1-\frac{i^2}{c^2}}\right) \qquad \text{4G8}$$

This is the created by fission *proton process demand* of the neutron collapse of the new elements. This energy is equal to the so called "*neutron radiation*", the proton process intensity intake.

So, the result is that what has been earned at one side of the process as *blue shift* intensity benefit – energy generation – the same has been lost on the other, on proton process cover and electron process *blue shift* need of the newly created elements. And all these have to be taken from other elements of the surrounding environment.

Proton process cover to be taken from the environment is not an easy case. Proton and electron processes have been connected. The new born elements need electron process drive of certain intensity with corresponding proton process cover. Otherwise neutron processes would stay without balance, turning into *gamma* radiation case.

Neutron radiation is not about flying neutrons.
(There has never been flying neutrons measured, as it was for *alpha* radiation.)
It is rather external proton process intensity cover and electron process *blue shift* drive intake in order to compensate the missing balance of the *Strong Interrelation*.

This is not about the neutron process "demand" of the new elements, with damaged elementary balance, products of the break of the *Uranium*.
The neutron process is passive.
This is about neutron processes remaining without sufficient drive and cover, results of the fission. The surrounding environment is the one, which is providing electron process *blue shift* drive and proton process cover to the neutron processes.
The quantum communication between the neutron processes, products of the fission and the environment results in elements, types of the "external support". At the same time the elements of the external support become isotopes with damaged balance.

Heavy elements or mixtures of minerals with neutron process intensity dominance as external support are less sensitive to this impact and can tolerate the loss of their partial *blue shift* drive and proton process intensity. These elements or mixtures can be used as shields. This permanent "shielding" function however is also with consequences and results in slow destruction of their elementary structure.

Neutron radiation has limited impact on water, but results in the generation of H_3O - *Tritium,* since water is losing on (unit) proton process intensity.

S.   **4.2**
4.2   **Consequences of *neutron* radiation**

"Neutron radiation" causes heavy damages within the elementary structures of the elements of the external support all around, resulting in *gamma* and *beta* radiation: The damaged elements lose on proton process *intensity* and electron process *blue shift* drive.

The "benefit" of the fission is energy generation in form of *blue shift* conflict (heat). This electron process *blue shift* surplus cannot be used for covering the damage, since the proton process cover is missing.

Fission products are all with damaged *Strong Interrelation*.
External elementary contribution helps in re-establishing the overall mass-energy balance but cannot compensate the damage.
There are isotopes, born as products of the break of the elementary process of the *Uranium* and isotopes, created as result of "neutron radiation".

The key of the neutron radiation is the loss on the working *capability* of the element of the external support. The supporting external "donor" element remains without sufficiently balanced neutron process.

4H1
$$\frac{dmc^2}{dt_p\varepsilon_p}\left(1-\sqrt{1-\frac{v^2}{c^2}}\right) = \frac{dmc^2}{dt_p\varepsilon_p^r}\left(1-\sqrt{1-\frac{v^2}{c^2}}\right) + \frac{dmc^2}{dt_p\varepsilon_p^{nf}}\left(1-\sqrt{1-\frac{v^2}{c^2}}\right)$$

ε_p^r - the remaining intensity of the proton process

ε_p^{nf} - the intensity taken by "neutron radiation"

The remaining capacity of the damaged (by "neutron radiation") element in proton process and electron process intensities is obviously less than their standard value. And the chain of damages starts.

Ref   *Gamma* radiation, reference to 4E3, is of high energy impact:
4E3
4H2
$$w_\gamma = \frac{dmc^2}{dt_i\sqrt{1-\frac{(i-u)^2}{c^2}}}\sqrt{1-\frac{(c-i)^2}{c^2}}\left(1-\sqrt{1-\frac{[c-(i-u)]^2}{c^2}}\right);$$

The high energy in 4H2 is not just about the value of the second bracket, which even in the worst scenario is "just" doubling the impact, rather the slow-down of the time flow (intensity increase) resulting in high frequency.

The resulting time count is: $dt_u = dt_i\sqrt{1-\frac{(i-u)^2}{c^2}}$ In the case of $u = i$ the *gamma* work is zero, the status is balanced
– it is electron process.

While *neutron* radiation directly causes *beta* and *gamma* type radiation, *gamma* may further impact and destroy balanced proton-neutron intensity relations of other elements and resulting in extended damage of either *alpha* or *beta* radiation.

4.3
Natural recovery from *isotope* damage

Natural recovery is the "self-rehabilitation" of the *Strong Interrelation* of isotopes.

The damaged intensity relation is:
$$\frac{dt_n^*}{dt_p^*} \neq \frac{\varepsilon_p}{\varepsilon_n}\sqrt{1-\frac{(c-i)^2}{c^2}} = \varepsilon_e$$

4I1

We take that there are η proton-electron-neutron process cycles within the damaged element. As a result of the damage, however, only $(\eta - x)$ of these processes corresponds to the standard *Strong Interrelation* of the element and will be reaching the inflexion point in their normal and balanced way and will start the new cycle with standard to the element proton process intensity.

$\eta - x$ are reaching the inflexion point and continue with normal intensity:

$$\text{(relation after)} \quad \frac{\eta - x}{\eta} + \frac{\eta - x}{\eta - x} > \frac{\eta - x}{\eta} \quad \text{(relation before)}$$

4I2

and the relations are improving.

Reaching the inflexion point is crucial, since this fact is a direct signal to the *Strong Interrelation*, the acting balance between the proton and the neutron processes. This takes time however, sometimes long years.

Natural recovery is obvious: it only needs time and the necessary number of cycles to happen. The lifetime of isotopes, therefore, can be calculated and the impact they cause via/to the *Quantum Membrane* is limited in time. The impact of isotopes is either dissipating within the *Quantum System of Reference*, or causing damage in other elementary structures.

Recovery with external support means *speeded up* rehabilitation of the original balance of the proton-neutron processes of the damaged element – by external support of other elements.

Are there any kinds of donor elements or elementary compositions, which can provide missing proton or neutron process intensity support?

Chemical reactions are based on balanced communication between elements, but simple chemical reactions will not improve the imbalanced status of isotopes. In the case of isotopes the balanced communication is limited and the new chemical composition would also be with damaged elementary structure.

S.   **4.4**
4.4   **Balance of *Strong* and *Weak Interrelations***

Chemical reactions in general are based on the intensity relations of the electron processes of elements and compounds.

If we take two elements or two compounds of elements, one with electron process deficit (proton process intensity deficit) and the other with electron process surplus (proton process intensity surplus), the two opposite parts will be compensating each other's demand and will establish the optimal balance of the elementary processes.

The mix of elements with proton process intensity and electron process surplus (donors) will provide *blue shift* to the other part (recipients) of the compound with the missing intensities of the proton and electron processes.

The intensity of the electron process, acting on the recipient element corresponds to the intensity of the electron process of the donors. The coverage of the neutron process of the recipient is coming from the proton process of the donor. The donor element is recycling using the neutron of the recipient element.

With the provision of electron process surplus and proton cover, naturally available within the donor, the donor provides additional drive and cover to the neutron collapse of the recipient.
Electron processes of the communicating compounds or elements will be naturally harmonised. Components with increased proton process intensity will be communicating with those of high neutron process intensity demand. On the other hand, the less intensive proton and neutron processes will also find each other. The harmony will be optimal, depending on the composition of elements.

As drive of the quantum communication, the more intensive processes will be dominating and leading the communication.
Elementary processes will be corresponding to the energy transfer between the donor and the recipient elements in line with the proton process and the neutron process intensities of the compounds.
Donors with dominant proton process intensity and electron process surplus will impact all available neutron processes, a function only of the capacity of the proton and electron process intensities of the donors.

If the elementary communication is about x elements and

4J1
$$\frac{dm}{dt}_{p1} > \frac{dm}{dt}_{p2} > \frac{dm}{dt}_{p3} > ... > \frac{dm}{dt}_{px}$$

all elements will be communicating, but elements with the highest value of proton and electron process intensities (in this particular case the proton and electron process intensities of element *1*) will dictate the neutron collapse of all others with uncovered neutron process intensity demand. But elementary processes are of infinite cycles, therefore, the other, less intensive proton and neutron processes of the elementary composition will also be communicating.

This means that elements with intensive proton processes and others with intensive neutron process demand can create a strong balance close to equilibrium. (Since more intensive proton process consequently has less intensive neutron process demand and vice versa). Elements with neutron process intensity dominance will have the chance to make "corrections" in the quality of the mixture, having additional electron process drive and proton cover in line with the actual proportions of the components. Communications happen in parallel at all intensity levels.

The more intensive and less intensive communications (reactions) between the components of the mix are always in harmony. The optimal balance will be built up at a certain point – including the "corrections" (adjustments in proton/neutron relations) of the capacities of the two sides.

The elementary communication follows the intensity drive of the dominant elements. This does not change the mass-energy transfer and the integrated capacity relations of the communication. This just regulates the process in a certain way. Elementary processes happen with higher integrated intensity for shorter time. Otherwise, without communication, the same transfer would happen with less intensity for longer time. The effect of the communication means an increase of intensity.

Element with the highest $\dfrac{dm}{dt}_{p\,\text{max}}$ proton process intensity 4J2

and with the highest surplus of electron process intensity:

$$\frac{dmc^2}{dt_i \varepsilon_e}\left(1 - \sqrt{1 - \frac{(c-i)^2}{c^2}}\right) \qquad 4\text{J}3$$

will have maximum impact on other elements:
It is important to note that as consequence of the *blue shift* drive and proton process cover impact, the neutron collapse after the inflexion point results in the same element as of the *impact*, if the continuity of the electron process *blue shift* impact is guaranteed and the proton cover is provided.

The benefit of the communication is that the proton process and the electron process intensities of the donor elements not just speed up the elementary processes of the recipients, but the benefit of the "energy" potential of elements with electron *blue shift* surplus also can be utilised.

4.5
Rehabilitation of *isotopes*

Isotopes are with damaged *Strong* and *Weak Interrelations*.
External quantum support can speed up the self-rehabilitation process.
It means shortening of the half-life time of isotopes.

The task is to have impact on the elementary processes of the isotopes and re-establish the balance.
The speeding up means, finding the necessary elementary composition for efficient quantum communication in order either to drive the neutron process or use the available electron process *blue shift* drive surplus and utilise the proton process dominance.

Elements supporting the rehabilitation either provide or take off quantum energy. Quantum energy is provided by the *blue shift* of the electron process and is realised by the increase of the intensity of the neutron collapse. Proton process after the inflexion point starts with increased energy intensity.

The recovery process ensures the isotope returns back to its normal balanced status.

β^- radiation means certain missing electron process *blue shift* drive and proton cover.
The deviation is far from *gamma* radiation status: the missing electron process intensity and proton process cover do not result in the turnover of the neutron collapse, but isotopes take electron process *blue shift* and proton process cover from the environment. The missing electron process *blue shift* drive and proton process cover to the neutron collapse of the isotope can be provided by donor elements with proton process dominance and electron *blue shift* surplus. As the donor element also gives the *proton process cover* to the neutron collapse, the intensity at the inflection point corresponds to the proton/neutron intensity relation of the donor element.

This way, the elementary process of the isotope will be accelerated:
The intensity of the electron process drive of the donor is forcing the process ahead with higher speed than that would follow from the elementary characteristics of the isotope itself. In the meantime, the existing and available normal *blue shift* drive of the isotope – even of less intensity – will be utilised (by the isotope itself or by others at the donor site) as part of the harmony principle of the communication.

The rehabilitation is more efficient with a mix of elements. The composition of a possible decontamination mix for speeding up the rehabilitation of β^- isotopes (like *Cs-137*, *Co-60* or *Sr-91*) may contain elements, close to the absolute balance status of their proton process intensities and neutron process intensity demands, like *O, N, C, S, Ca, Si* from the side of the proton process intensity and *Mg, Al, K, P, Cl, Ti* from the side of the neutron process intensity demand.

Proton and neutron processes of the "decontamination" mix are in balance in absolute terms:

$$\frac{dmc^2}{dt_p^a \varepsilon_p^a}\left(1 - \sqrt{1 - \frac{i^2}{c^2}}\right) + \ldots + \frac{dmc^2}{dt_p^x \varepsilon_p^x}\left(1 - \sqrt{1 - \frac{i^2}{c^2}}\right) =$$

$$= \frac{dmc^2}{dt_n^a \varepsilon_n^a}\xi_a\sqrt{1 - \frac{(c-i)^2}{c^2}}\left(1 - \sqrt{1 - \frac{i^2}{c^2}}\right) + \ldots + \frac{dmc^2}{dt_n^x \varepsilon_n^x}\xi_x\sqrt{1 - \frac{(c-i)^2}{c^2}}\left(1 - \sqrt{1 - \frac{i^2}{c^2}}\right)$$

4K1

The intensity balance of the decontamination mix is close to equilibrium:

$$\frac{dmc^2}{dt_p^a}\left(1 - \sqrt{1 - \frac{i^2}{c^2}}\right) + \ldots + \frac{dmc^2}{dt_p^x}\left(1 - \sqrt{1 - \frac{i^2}{c^2}}\right) \approx$$

$$\approx \frac{dmc^2}{dt_n^a}\sqrt{1 - \frac{(c-i)^2}{c^2}}\left(1 - \sqrt{1 - \frac{i^2}{c^2}}\right) + \ldots + \frac{dmc^2}{dt_n^x}\sqrt{1 - \frac{(c-i)^2}{c^2}}\left(1 - \sqrt{1 - \frac{i^2}{c^2}}\right)$$

4K2

The *absolute balance* of the isotope however is *broken*:

$$\frac{dmc^2}{dt_p^i \varepsilon_p^i}\left(1 - \sqrt{1 - \frac{i^2}{c^2}}\right) \neq \frac{dmc^2}{dt_n^i \varepsilon_n^i}\xi_i\sqrt{1 - \frac{(c-i)^2}{c^2}}\left(1 - \sqrt{1 - \frac{i^2}{c^2}}\right)$$

4K3

Once the decontamination mix, the composition of the elements above in integrated quasi balance is mixed with the isotope, all elements with electron process *blue shift* surplus and proton process intensity dominance will be communicating with the isotope, as the isotope is in need of both.

The communicating elements have *blue shift* and proton process intensity surplus. Being in quantum communication with the neutron process of the isotope means driving the neutron process of the isotope into collapse and covering it by proton process intensity.

For introducing the rehabilitation process on a certain example, it is important to note that the standard intensity demand of the neutron process of the damaged *Caeseum-137* isotope is 1.4 times more than the intensity of its proton process.

The proton process intensity of the normal *Sulphur*, one of the components of the mix is 1.01 times more than its neutron process intensity demand. In *Aluminium* – another element of the mix, the neutron process intensity demand is more than its proton process intensity (as having an integrated balance of the mix) – but the intensity relation of the proton process and the neutron process demand is close to equilibrium: 0.94.

These figures are to demonstrate the capacity of the donor elements for speeding up elementary communication. Proton process dominant elements will provide electron process *blue shift* and proton process cover to the isotope. And with that the balance of the isotope starts to be re-established.

As consequence of the support in rehabilitation the used *blue shift* drive and proton process cover is missing from the elementary process of the donor elements of the decontamination mix. The earlier absolute balance, now with the rehabilitation of the isotope going on is broken:

$$\frac{dmc^2}{dt_p^a \varepsilon_p^a}\left(1 - \sqrt{1 - \frac{i^2}{c^2}}\right) + \ldots + \frac{dmc^2}{dt_p^x \varepsilon_p^x}\left(1 - \sqrt{1 - \frac{i^2}{c^2}}\right) + \frac{dmc^2}{dt_p^i \varepsilon_p^i}\left(1 - \sqrt{1 - \frac{i^2}{c^2}}\right) \neq$$

4K4

$$\neq \frac{dmc^2}{dt_n^a \varepsilon_n^a}\xi_a\sqrt{1 - \frac{(c-i)^2}{c^2}}\left(1 - \sqrt{1 - \frac{i^2}{c^2}}\right) + \ldots + \frac{dmc^2}{dt_n^x \varepsilon_n^x}\xi_x\sqrt{1 - \frac{(c-i)^2}{c^2}}\left(1 - \sqrt{1 - \frac{i^2}{c^2}}\right) +$$

$$+ \frac{dmc^2}{dt_n^i \varepsilon_n^i}\xi_i\sqrt{1 - \frac{(c-i)^2}{c^2}}\left(1 - \sqrt{1 - \frac{i^2}{c^2}}\right)$$

The benefit of the decontamination and the quantum communication, however, is that the *self-rehabilitation process* of the new born "donor-isotopes" within the decontamination mix, their half-lifetime is much shorter. It is within the region of seconds and minutes. This way the rehabilitation of long lived and dangerous isotopes can be shortened.

Part of the isotope element disappears, since the electron process drive and the proton process cover determine the element. Meaning, the rehabilitated neutron process of the isotope results in the donor element. This is not an *alchemic* result, rather the proof of the unity of the quantum processes.
There have not been produced new elements. The neutron process is passive and the *blue shift* drive and the proton process cover determine the quality of the element.

β^+ means increased electron process intensity. The isotope gives off electron process *blue shift* drive and proton process cover.
In the case of β^+ isotope with *blue shift* impact surplus and proton cover, the donor elements of the mix will be driving the neutron collapse of the isotope by their less intensive electron process *blue shift* drive and proton process cover.

At the same time the isotope will be forced to provide its available *blue shift* drive and *proton process intensity cover* to the element of the mix with increased neutron process demand.
This way the increased electron *blue shift* surplus and proton cover will be utilised. The new born isotopes have shorter half-life time.

The principle of the rehabilitation of a*lpha* isotopes would be similar to that of the β^+ isotope case; the rehabilitation of *gamma* to the β^- isotope case.

The rehabilitation of these damages, however, would need high volume of mix, as the surplus (*alpha*) and the miss (*gamma*) of proton processes are significant. Therefore, this so called "donor based" decontamination is efficient and doable mainly for *beta* isotopes.

4.6

Hydrogen – Helium blue shift* conflict rather than *fusion

S.
4.6

The electron process is:

$$\frac{dmc^2}{dt_i \varepsilon_i}\left(1-\sqrt{1-\frac{(c-i)^2}{c^2}}\right)=\frac{dn}{dt_i \varepsilon_i}q \qquad \text{4L1}$$

$$\overset{He \qquad\quad H}{\diagup\;\diagdown}$$

$$\varepsilon_{He}=X \qquad \lim\varepsilon_H=0$$

The difference in neutron process intensities: $\dfrac{dm}{dt_n^H} << \dfrac{dm}{dt_n^{He}}$ 4L2

The utilised work values of the electron process *blue shift* impacts within these two elements are completely different.

The intensity of the electron process within the *Hydrogen* in fact remains without use – the neutron process is of infinite low intensity.

In the infinite cycle of mass-energy-mass… transformation (of the matter) at its *Helium* stage the electron process *blue shift* impact as drive of the neutron process is acting with intensity surplus.

Contrary to the *fission process*, the *fusion* of *Hydrogen* atoms can never create *Helium*, since the neutron process of the *Helium* is of infinite times more intensive than that of the *Hydrogen*.

In the case of *fission*, with reference to 4G4, the less intensive neutron process of the newly born fission products makes it possible to utilise the electron process intensity surplus of the original element. The process follows the natural sequence of elementary chain: more intensive neutron processes are followed by less intensive ones.

Ref
4G4

Any *fusion idea* of *Hydrogen* elements into *Helium* is impossible.

This is not just against the natural process – as from a less intensive collapse cannot follow a more intensive one – but *Hydrogen* does not have the necessary *blue shift* impact intensity the *Helium* needs for its neutron collapse.

Hydrogen is with infinite volume of *blue shift* surplus

4L3

$$\lim \frac{dt_n}{dt_p} = \lim \frac{\varepsilon_p}{\varepsilon_n} \sqrt{1 - \frac{(c-i)^2}{c^2}} = \lim \varepsilon_x = 0$$

but the intensity of the electron process is *infinite low*.

The *plasma status* is not about the transformation of *Hydrogen* into *Helium*, rather about electron process *blue shift* conflict between elements with electron processes intensity surplus. The *plasma status* is *blue shift* conflict of infinite intensity.

5
Elementary communications

The balance of proton and neutron processes with electron process drive is the basis of elementary relations.

The internal process of elements with proton process dominance and electron process *blue shift* surplus is fully understandable.

Neutron process dominance is less obvious, since in this case the neutron process is more intensive than the proton process while the process goes with electron process *blue shift* "deficit".

In fact, this is not really about electron process deficit. We can call it this way, but the case is different: It means the neutron process uses all electron process *blue shift* available. The time system of the neutron process is "shorter" than the proton process. This way it is more intensive.

$$\frac{dmc^2}{dt_p \varepsilon_p}\left(1 - \sqrt{1 - \frac{i^2}{c^2}}\right) = \frac{dmc^2}{dt_n \varepsilon_n}\xi\sqrt{1 - \frac{(c-i)^2}{c^2}}\left(1 - \sqrt{1 - \frac{i^2}{c^2}}\right)$$

5A1

$$\varepsilon_e = \frac{\varepsilon_p}{\varepsilon_n}\sqrt{1 - \frac{(c-i)^2}{c^2}} \quad \text{(as it is about a relation = no dimension indeed.)}$$

5A2

In the case of neutron intensive elements (as $\varepsilon_n > \varepsilon_p$): $\varepsilon_e < 1$

[ε_e is the intensity coefficient of the electron process, characterising (adjusting) the *mass change* of the electron process within the standard time system of $i = \lim a\Delta t = c$; ε_i is representing an intensity value, related to the constant dt_i time system at $i = \lim a\Delta t = c$ speed.]

How could drive a less intensive electron process a more intensive neutron process in neutron process dominant elements?
Is the electron process in neutron dominant elements really less intensive?

The time system of the electron processes is constant: $i = \lim a\Delta t = c$ for all elements. This common time system of electron processes gives the chance to their direct communication. The description of the work formula of the electron process in absolute terms, with reference to 4D8 is:

Ref 4D8

$$W_{electron} = \frac{dmc^2}{dt_i \varepsilon_i \varepsilon_e}\left(1 - \sqrt{1 - \frac{(c-i)^2}{c^2}}\right);$$

5A3

The intensity formula of the electron process in this case will be:

5A4
$$e_{electron} = \frac{dmc^2}{dt_i \varepsilon_e}\left(1 - \sqrt{1 - \frac{(c-i)^2}{c^2}}\right); \qquad \text{as } \varepsilon_e < 1$$

which gives in fact *increased electron process energy (or work) intensity value* as it is expected for a more intensive neutron process! t_i - corresponds to $i = \lim a\Delta t = c$, equal and quasi the same for all electron processes; ε_e- is the distinguishing intensity coefficient of the electron process, reference to 5A2 addressing the proton/neutron intensity relation of the element at constant time system of t_i.

Ref
5A2

In the case of proton process intensive elements, the characterisation has to be given in the opposite way: electron process will be of less intensity, (while with *blue shift* surplus) since with reference to 5A2 $\varepsilon_e > 1$: 5A4 will be resulting *in electron process of less energy (or work) intensity value*. The neutron process in these elements is driven by electron process of less intensity.

Ref
5A2

> Elements with proton process dominance have more intensive transformation from mass into energy, which is equal to less intensive re-transformation from energy to mass – this is the explanation of the less weight value. And these elements are with electron process *blue shift* surplus and conflict. Elements with neutron process dominance have more intensive re-transformation of energy into mass, with certain electron process deficit but of increased intensity and higher elementary weight.

Strong Interrelation is the key of the communication between elementary processes.

In the case of *proton process intensive* elements the proton process needs less time to happen: $dt_p < dt_n$ since

5A5
$$\frac{dmc^2}{dt_p}\left(1 - \sqrt{1 - \frac{i^2}{c^2}}\right) > \frac{dmc^2}{dt_n}\sqrt{1 - \frac{(c-i)^2}{c^2}}\left(1 - \sqrt{1 - \frac{i^2}{c^2}}\right)$$

As the neutron process of this kind of elements is more time consuming (as being less intensive) there is a certain shift in the duration of the two processes, which will be resulting in electron *blue shift* surplus.

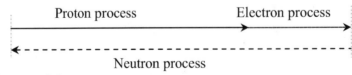

Diag.
5.1   Diagram 5.1

In the case of *neutron process intensive* elements

5A6 $$\frac{dmc^2}{dt_p}\left(1 - \sqrt{1 - \frac{i^2}{c^2}}\right) < \frac{dmc^2}{dt_n}\sqrt{1 - \frac{(c-i)^2}{c^2}}\left(1 - \sqrt{1 - \frac{i^2}{c^2}}\right)$$

the time shift is the opposite.

More intensive neutron process needs shorter time period: $dt_p > dt_n$

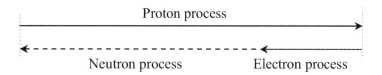

Diag.

Diagram.5.2   5.2

It is important to note that the absolute mass-energy values of the proton and neutron processes (with reference to the natural *quantum* deviation of each cycle = entropy) will always be equal to each other. Whatever are the durations of these processes new proton process and new elementary cycle only starts once the neutron process reaches the inflexion point.

With infinite number of elementary cycles of each element, this inflexion condition is not a problem, as proton, electron and neutron processes of different cycles communicate. Electron processes drive the neutron collapse and the energy of the proton process covers the mass re-creation of the neutron process.

In the case of isotopes this certain energy-mass-energy-mass… balance has been broken. The cycle of the elementary process goes with missing full proton-neutron cover, causing miss-communication.

In the case of $(n-1)$, *infinite number* of continuous and simultaneous elementary processes electron processes provide either *blue shift* "surplus" or work with *blue shift* "deficit" within the communication.

In the case of infinite numbers of elementary relations, the electron process *blue shift* surplus or deficit or the proton process versus neutron process dominance does no result in process continuity problems. The infinite number of proton/neutron relations always ensures the full correspondence.

Elementary communication between elements results in various characters of relations:

Element A

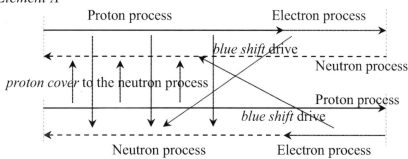

Element B

Diag.

Diagram 5.3   5.3

Communication between elements means *utilisation of* the electron process *blue shift* surplus by other elements having electron process *blue shift* deficit. The overall balance equalises the *blue shift* surplus and the deficit.

In other words, elements with *blue shift* surplus simply "offer" their electron process capacity for utilisation.

Ref.
Diag.
5.3

Diagram 5.3 demonstrates this kind of communication in simplified way.

As result, communication of A and B will be in harmony and full balance.
Element B with neutron process dominance and electron process *blue shift* deficit utilises the *blue shift* surplus of Element A. At the same time the proton process of Element B will be covering the neutron process of Element A under the control of the necessary electron *blue shift* reserve of Element B.

Ref.
Diag.
5.3

In the communication above in Diagram 5.3 we can distinguish *three* parallel segments with different intensities:
-   (1) alongside the cover of the obvious internal *blue shift* demand of each of the components as *one* of the *segments*;
-   (2) the communication will be utilising the *blue shift* surplus of Element A (of proton process dominance) in the *segment* of the neutron collapse of Element B;
-   (3) the *other segment* is the generating *blue shift* of Element B, to be utilised for the "used" electron process intensity of Element A for its less intensive neutron collapse.

Elements A and B, being in balance before separately will also be after it: in "more efficient" way, using each-other's potential.

One important aspect shall be taken always into account:
Available electron process *blue shift* can only be utilised by neutron process if the proton process cover is also available! If not, the *blue shift* available results in conflict or/and increased intensity, rather than elementary communication.

Ref.
Diag.
5.3

The case in Diagram 5.3 is idealistic, since the two elementary processes are in proton/neutron process symmetry to each other. This kind of relations may results in full, *unbreakable* structure and balance.

Relations in practice are more realistic as communications are far not in proton/neutron process symmetry. Communication corresponds to the relation of electron process intensities of the components.

In the case of limited number of elements, the communication can be distinguished in time as well:
The communication at the highest intensity level (as per segment 2 above) can be identified as the one happening for the shortest period of time.
The *following*, less intensive segment/s need longer time *period/s*.

This way, reaction or communication between elements can be distinguished for periods, while they are happening in infinite numbers in fact in parallel.

The first period is always about the most intensive proton-neutron communication. Later stages usually "resolve" strong connections for certain extent established during the first the period/s.

Elementary processes remain processes and components will be recreated during communication. Periods in fact do not have ends. Intensities are changing in each cycle. The formulated in the first period/s "unbreakable" structure/s either will be resolved by still acting in the mix (with less intensity) *blue shift* surplus of one (or number) of the components; or will be soften by the still existing and dominant within the mix *blue shift* deficit of one (or number) of components.

As conclusion: the electron process drive and the proton process cover determine the element. Independently, what is the type of the element with electron process deficit, a neutron process driven by electron process *blue shift* and covered by its respective proton process will always generate element of the *blue shift* drive and the proton process cover.

5.1
Increased elementary communications, principle of rehabilitation of isotopes

S.
5.1

Ideal and "unbreakable" proton-neutron intensity balance is:

$$\frac{dmc^2}{dt_p}\left(1-\sqrt{1-\frac{i^2}{c^2}}\right) \cong \frac{dmc^2}{dt_n}\sqrt{1-\frac{(c-i)^2}{c^2}}\left(1-\sqrt{1-\frac{i^2}{c^2}}\right)$$

5B1

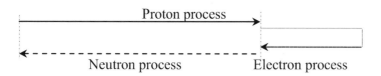

Diagram.5.4

Diag.
5.4

This is not just about the absolute balance in 5B1, but also about the quasi intensity balance as well.

Balanced elementary structures however do not need the equality of the intensities of proton and neutron processes. Balanced elementary structures mean the absolute balance of the proton and neutron processes:

$$\frac{dmc^2}{dt_p\varepsilon_p}\left(1-\sqrt{1-\frac{i^2}{c^2}}\right) = \frac{dmc^2}{dt_n\varepsilon_n}\xi\sqrt{1-\frac{(c-i)^2}{c^2}}\left(1-\sqrt{1-\frac{i^2}{c^2}}\right)$$

5B2

Isotopes in the contrary, mean destroyed absolute balance:

5B3
$$\frac{dmc^2}{dt_p\varepsilon_p}\left(1-\sqrt{1-\frac{i^2}{c^2}}\right) \neq \frac{dmc^2}{dt_n\varepsilon_n}\xi\sqrt{1-\frac{(c-i)^2}{c^2}}\left(1-\sqrt{1-\frac{i^2}{c^2}}\right)$$

In the case of β^- (electron) radiation, the neutron process of the damaged element is short of the necessary electron process *blue shift* drive and proton process cover. For covering the demand, isotope takes electron process *blue shift* and proton process cover from the environment without reciprocal balance of the elementary communication.

In the case of β^+ (positron) radiation, the element has extra *blue shift* impact, because of the damaged (increased) proton/neutron process intensity relation.

Should the missing electron process *blue shift* impact and proton process cover be provided (β^-) or utilised (β^+) the broken absolute balance could be re-established.

In the case of a certain composition of number of elements with integrated proton process dominance and electron process *blue shift* surplus in absolute balance (as per 5B4) and in intensities (as per 5B5 below),

5B4
$$\sum\frac{dmc^2}{dt_p\varepsilon_p}\left(1-\sqrt{1-\frac{i^2}{c^2}}\right) = \sum\frac{dmc^2}{dt_n\varepsilon_n}\xi\sqrt{1-\frac{(c-i)^2}{c^2}}\left(1-\sqrt{1-\frac{i^2}{c^2}}\right)\;;$$

the available *blue shift* surplus and proton process cover may help to drive and cover the neutron process of the isotope (β^-), which otherwise is missing its sufficient *blue shift* drive and proton process cover.

5B5
$$\sum\frac{dmc^2}{dt_p}\left(1-\sqrt{1-\frac{i^2}{c^2}}\right) > \sum\frac{dmc^2}{dt_n}\sqrt{1-\frac{(c-i)^2}{c^2}}\left(1-\sqrt{1-\frac{i^2}{c^2}}\right)$$

5B5 demonstrates: the integrated value of proton process intensities is more than the integrated intensity of the neutron processes.

The diagram of the relation is:

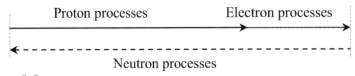

Diag.
5.5   Diagram 5.5

Once the neutron collapse of an isotope reaches the inflexion point the newly born proton process will start a "healthy cycle".

In the case of a mix with elements with potential electron process *blue shift* surplus, proton-neutron relations within the mix will follow the neutron process demand in line with the available electron process intensity drive and proton process cover of the components.

Electron process *blue shift* surplus can be achieved within a mix if elements close to equilibrium status have been selected in certain composition.

In line with the principle shown in Diagram 5.3 the electron process *blue shift* drive of proton process intensive elements will strengthen the drive of the neutron processes of all other elements. As outcome, there could be different options, but the final result in any circumstances will be an integrated *Strong Interrelation* balance optimum with the most efficient use of all electron process *blue shift* and proton process cover available.

Ref. Diag. 5.3

Proton process intensive elements have neutron processes of less intensity. The neutron processes of these proton process dominant elements utilise *blue shift* drive available at later stages of the communication. Electron process *blue shift* drive for these neutrons will be provided during the less intensive segments of the communication by the *blue shift* of those elements, which neutron processes have already been driven by proton intensive elements.

Once a β^- isotope is included into a mix of elements, the balance process of the mix will be different. As response to the natural electron process drive need and proton process cover demand of the isotope, its neutron process will be driven and covered by electron process *blue shift* and proton process cover available within the mix.

All proton process dominant elements will provide *blue shift* to their own neutron collapse and to the neutron collapse of all other elements with neutron dominance, including also the *isotope*.
This way the neutron process of the *isotope* will be driven.

In parallel all neutron process dominant elements provide *blue shift* to their own neutrons including the *isotope* as well (in the volume the demand alongside the more intensive external contribution makes it necessary).

The elementary process of the *isotope* is in comfort, since its neutron process is fully driven by the two (external and internal) segments. The used – for the rehabilitation of the isotope – external electron process *blue shift* and proton process cover however will be missing now from the balance of the donor elements of the mix.

The absolute balance of the integrated *Strong Interrelation* of the mix can be written as:

$$\sum (1+x)\frac{dmc^2}{dt_p\varepsilon_p}\left(1-\sqrt{1-\frac{i^2}{c^2}}\right)=\sum \frac{dmc^2}{dt_n\varepsilon_n}\xi\sqrt{1-\frac{(c-i)^2}{c^2}}\left(1-\sqrt{1-\frac{i^2}{c^2}}\right) \qquad 5C1$$

Since the proton process intensity of the *isotope* is damaged (weakened), other components of the mix have to provide the *blue shift* for managing the balance. This is the meaning of the increased $(1+x)$ proton portion.

$$(1+x)\frac{dmc^2}{dt_p}=\frac{dmc^2}{\dfrac{dt_p}{1+x}}$$

It means integrated utilisation of *more* proton process intensities of the mixture – including the isotope – for keeping the balance.

5C1 is with a deliberate correction otherwise the integrated absolute balance of the mixture with the isotope also would be in non-balance.

5C2 $$\frac{dm}{dt_p}(1+x)$$ The correction is that this integrated mass change intensity does not correspond to the integrated standard intensity ε_p of the elements of the mix.

But 5C1 could be written in another way as well, equivalent to

5C3
$$\sum \frac{dmc^2}{dt_p \frac{\varepsilon_p}{1+x}}\left(1-\sqrt{1-\frac{i^2}{c^2}}\right)=\sum \frac{dmc^2}{dt_n \varepsilon_n}\xi \sqrt{1-\frac{(c-i)^2}{c^2}}\left(1-\sqrt{1-\frac{i^2}{c^2}}\right)$$

5C4 $$\varepsilon_p^x = \frac{\varepsilon_p}{1+x}$$ 5C3 this way demonstrates the less integrated virtual proton process intensity value of the mix with the isotope.

At the same time dt_p the time frame of the proton process corresponds to more intensive mass change.

Ref
5C1
5C3

There is no difference what is the interpretation or which equation we use either 5C1 or 5C3, the result, regarding to the electron process intensity of the mix will be the same:

ε_e^x - the specific characteristic of the

5C5   electron process of the mix, $$\varepsilon_e^x = \frac{\varepsilon_p}{(1+x)\varepsilon_n}\sqrt{1-\frac{(c-i)^2}{c^2}}$$

which, with reference to 5A2, indicates electron process with more intensive mass change, since

5C6
$$\frac{dmc^2}{dt_i \varepsilon_i \varepsilon_e}\left(1-\sqrt{1-\frac{(c-i)^2}{c^2}}\right) < \frac{dmc^2}{dt_i \varepsilon_i \varepsilon_e^x}\left(1-\sqrt{1-\frac{(c-i)^2}{c^2}}\right)$$

This way the missing *intensity* of the mix because of the β^- isotope is compensated by certain increased *work load*, by more intensive integrated electron process: utilisation of more electron process *blue shift* from the proton process intensive elements of the mix.

When the neutron process of the isotope is reaching the inflection point the damage in the balance is resolved. The "new" proton cycle will be normal, corresponding to the donor element rather than to the isotope.

The objective of isotope rehabilitation by mix of proton process dominant elements is to compensate the missing electron process *blue shift* drive and proton process cover of the isotope. *Blue shift* portion is taken from the proton process dominant part of the mix and also proton process is provided for covering the neutron process demand of the isotope. (This is part of the more intensive segment of their elementary communication.)

With the rehabilitation however there will be still no absolute balance within the mix!

At certain stages of the elementary communication of the mix, proton process dominant elements will be the ones which will be missing the electron process *blue shift* drive, as their own electron process *blue shift* drive has already been used for the rehabilitation of the original isotope.

The original *isotope* has been rehabilitated for the price of "creating" new *isotope/s*.

The *benefit* of this rehabilitation process is that the damage of the elementary process balance of the original isotope is moved to other elements. These become isotopes indeed, but with significantly shorter half-life (self-rehabilitation).

5.2

Rehabilitation by single element and mix

The broken proton-neutron *Strong Interrelation* of an isotope with electron process intensity and proton process cover demand (case of β^-) might be resolved if mixed with another element with electron process surplus.

The donor element with electron process intensity surplus and proton process cover will provide the necessary drive and cover to the neutron process of the isotope. But this drive and cover will be missing from the elementary balance of the donor element at the end of its rehabilitation.

Elementary structures are built up on the intensities of the electron process and the proton process cover. Any neutron collapse can be driven by any electron process *blue shift* if proton cover is available. The intensity of the electron process, followed by the necessary proton process cover determines the intensity of the inflexion point, the most important characteristic, the "birth" of the element (the start of the proton process).

If the neutron process of the isotope has been fully driven and fully covered by electron and proton processes of the donor element, the element in fact is changed. But electron process intensities, used for the rehabilitation of the isotope will be missing from the balance of the donor element. This will be resulting in an *isotope* of the donor element.

If the electron process need and proton process cover demand of the neutron process of the isotope has not been fully covered, the isotope will still be there, plus a new one will be appearing in addition, belonging to the donor element.

Both results are equivalent to resolving isotope concentration: The original isotope either disappears or still exists but of less concentration. There will be however a new isotope in the scope – but very likely of less intensity – as the donor element has higher energy reserve.

If isotope is surrounded by elements with high energy content close to equilibrium status, the isotope will be under double effect:

(1) elements with electron process surplus will be driving the neutron process of the isotope;

(2) elements with neutron process dominance but close to proton/neutron process equilibrium state, first will also be utilising electron *blue shift* surplus, but later will provide electron process intensity and proton cover to the isotope. And so, will extend the impact on the isotope.

Hydrogen, Oxygen, Nitrogen and *Helium* cannot be in single usable format as their existing *blue shift* impact is conflicting (making them gaseous) and of low intensity. All *blue shift* surplus of *Oxygen* is utilised by the neutron process of the *Hydrogen* in *water* and the electron process *blue shift* surplus of the *Hydrogen* is of infinite low intensity. Water as the benefit of its *blue shift* conflict is dissolving rather than rehabilitating. It is difficult to find clean *Coal, Calcium, Silicon* and *Sulphur* in the nature. Natural minerals connected to these elements utilise part of their existing *blue shift* surplus. Elements with slight neutron dominance like *Al, Mg, K, Na, Cl, P, Ni, Fl, Ti, Va, Sc* make the balance with these elements stable and strong.

Minerals of *C, Ca, Si, S*, different *Oxides*, compounds of *Hydrogen* in composition with *Al, Mg, K, Na, Cl, P, Ni, Fl, Ti, Va, Sc* give a stable, balanced elementary structure close to equilibrium state, still with proton process dominance, ideal composition for mixtures for rehabilitating isotopes.

Priority cycles of the rehabilitation:
The electron process *blue shift* surplus of donor elements will drive the neutron process of all neutron process dominant elements including the *isotope* to collapse, providing the cover from proton process dominance elements of the mix. This will generate stable elementary structures of high stability.
Neutron dominance and isotope character will be still existing but in less proportions.
In other words, elementary processes of all elements of the mix (including the isotope) will be going on in parallel but the electron process surplus and the proton cover of the proton process dominant elements will create the most optimal elementary balance structure related to a limited portion of the mix.

The *following, later cycles* of the rehabilitation relate to the process when the electron process intensity surplus and proton process cover of the donor elements has already been partially and fully utilised. It does not mean these donor elements disappear – in the contrary these elements have been recreating from the driven and covered neutron processes of the isotope – just the distinguishing feature of their proton process dominancy is over.

The *blue shift* of the electron process of neutron dominant elements will continue acting, providing also proton process cover, but of less intensity.

As result, neutron process dominant elements will be recreating including the element of the isotope as well. The isotope will have back its balanced structure, as its neutron process will be driven to full collapse (to the inflexion point).

As final result the original *isotope* will be in balanced *cycle.*
At the same time other elements of the composition – proton and neutron dominant ones as well – will be with balance problems, as their proton process cover and electron process intensities will be of less value as have been used for the rehabilitation of the isotope. But as benefit, the half-life of these isotopes will be of short period.

5.3

Intensities of electron, neutron and proton processes

The relation of proton and neutron process intensities, the electron process intensity coefficient is the distinguishing characteristic of elements. Each element has its specific electron process intensity coefficient value.

$$\varepsilon_x = \frac{\varepsilon_p}{\varepsilon_n}\sqrt{1-\frac{(c-i)^2}{c^2}}$$

5D1

Electron process is the rule and the drive; proton process is the energy and neutron process is the work. The neutron process is the passive component. Electron process intensity and proton cover are establishing the element.

There are two intensity factors of the electron process:
ε_i - the one relating to the constant $i = \lim a\Delta t = c$ speed; and

ε_x - the intensity coefficient, relating to a certain element;

The radial sphere symmetrical expanding acceleration in the direction from the centre towards the periphery is the intensity of the process.
The *Quantum System of Reference* becomes loaded by the *blue shift* impact of the electron process and the *Quantum Membrane* transfers the impact as quantum communication.

The *radial acceleration* of the rotation depends on the status of the sphere symmetrical expanding acceleration.

The radial acceleration is: $a_r = \dfrac{dv}{dt} = \dfrac{v}{2\Pi}n$; where   5D2

Peripheral speed: Radial speed: and $v = 2\Pi \cdot \upsilon$

$v = \dfrac{2dR\Pi}{dt}$ $\upsilon = \dfrac{dR}{dt}$ or in its other format: $\upsilon = \dfrac{v}{2\Pi}$ 5D3

With the progress of the sphere symmetrical expanding acceleration of the electron process, the first component of the equation in 5D2 will be without change, remaining the same. But the spin number is decreasing as with the increase of the *radius* of the expansion and the duration grows to infinity.

-   The *radial acceleration* (intensity) of the electron process at the end stage is approaching: $\lim a_{er} = 0$

-   The neutron process starts by mass status having been expanded to infinity and acceleration, value of $\lim a_{rn} = 0$.

    This corresponds to the end status of the electron process.

-   The neutron collapse comes to its end at the inflexion point with acceleration, value of $\lim a_{rn} = \infty$

    This status is the starting point of the proton process, the sphere symmetrical expanding acceleration up to $i = \lim a\Delta t = c$.

-   The radial value of the sphere symmetrical expanding acceleration of the proton process at the start is: $\lim a_{rp} = \infty$.

-   The inflexion point is connecting neutron and proton processes, at the level of infinite value of intensity.

-   The end value of the radial acceleration of the proton process is equal to the starting acceleration value of the electron process at $i = \lim a\Delta t = c$, value of $a_{ip} = a_{ier}$

Proton and neutron processes also should mean rotation, or circulation at specific peripheral speed value.

Neutron process is collapse with the growth of the intensity of the collapse as the electron process *blue shift* impact is dictating in its growth and as the proton cover is slowing down at the end of its process, approaching $i = \lim a\Delta t = c$.

With this tendency, neutron is collapsing with increasing intensity (acceleration) and after the inflexion point the proton process starts with the intensity of full "power" of the collapse.

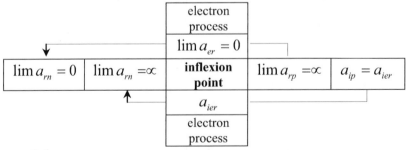

Diag.
5.4   Diagram 5.4

With reference to Diagram 5.4 proton and neutron processes are symmetrical, but with opposite intensity gradients.

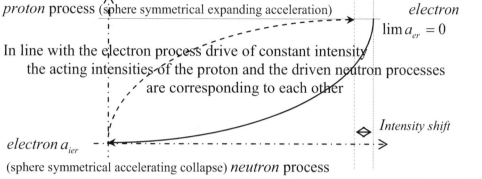

Diag.
5.5

Diagram 5.5

The proton process starts with *infinite high* intensity and continues with intensity of decreasing tendency. The neutron process is driven at the start by intensity of *infinite low* value but it continues by increasing gradient.

The end stage of the neutron collapse is of infinite intensity value, as mass is reaching its limits in collapse, while the intensity of the proton process is approaching zero, as it is close to the electron stage.

This is in full correspondence with the time systems of the processes.

The intensity of the electron process is constant, characteristic of the element:

$$\varepsilon_x = \frac{\varepsilon_p}{\varepsilon_n}\sqrt{1-\frac{(c-i)^2}{c^2}} = \frac{dt_n}{dt_p}$$

5D4

Acceleration values within 5D2 and 5D3 and on Diagrams 5.4 and 5.5 correspond to intensities.

- Infinite intensity of the proton process relates to $\lim a_{rn} = 0$ acceleration of the neutron process;
- Infinite intensity of the neutron process relates to the less intensive expanding value of the proton process.
- Electron process intensities vary between $\lim a_{er} = 0$ the end status of the electron process, relating to the drive at the start of the neutron collapse; and a_{ier} the start of the electron process, relating to the end stage of the neutron collapse with $\lim a_{rn} = \infty$ acceleration.
- Electron process, proton and neutron processes happen in time and space. In a case of $\varepsilon_x = \varepsilon_r = 1 = (a_r)$ proton and neutron processes happen in symmetrical way, one is expanding the other is collapsing.

The directions of the processes and the gradients of the intensities of the mass and energy exchange compensate each other. The resulting structure intends to be a non-breakable combination of generations of the element.

proton	Electron	Neutron	proton	electron
electron	Neutron	Proton	electron	neutron
neutron	Proton	Electron	neutron	proton
proton	Electron	Neutron	proton	electron

Diag.
5.6   Diagram 5.6

With reference to Diag.5.6, proton, electron, neutron processes represent always of three generations: each column is about simultaneous processes going in parallel; each line represents the continuity of the same process.

Actual elementary communications usually are not about the same cycle. *Proton process*, covering the neutron process of the element, driven by electron process represent the following three cycles:
Electron process cannot initiate the proton process of the same cycle, since electron process starts when the proton process is completed.
Neutron process cannot be driven within the same cycle, since the end stage of the electron process is the start of the neutron process.
This way, elementary processes of an existing element combines at least 3 following cycles.

This situation means that even the internal processes of the element with close to $\varepsilon_x = 1$ are not homogenous. It is result of simultaneous and parallel processes of three different generations (cycles).

S.
5.4

5.4
Inflexion point

With the growth of the periodic number the energy intensity of the electron process is growing!

In the case the proton process is of higher intensity, the neutron process is longer and the electron process is also of less energy intensity.
> This is the reason in this case of the low intensity value of the neutron process. The proton process is more active; mass-energy transformation is producing a surplus in electron process, fully acting in the element; but the time system of the neutron and proton processes are different.

The benefit of chemical reactions is that elements are communicating in order to reach a balanced status closer to equilibrium. Combination of elements with natural electron process surplus and natural electron process deficit stimulates chemical reaction between the two, using surplus to compensate deficit.
Cause and consequence are the same here.
In the case of an isotope the natural balance has been destroyed.

Beta means: elementary process goes on, but the proton process generates either more or less energy. In the case of less generation, the isotope takes electron process *blue shift* impact from aside, in the case of more generation the isotope gives off electron process *blue shift* impact.

The absolute work of the electron processes is:

$$\frac{1}{\varepsilon_e^i}\frac{dmc^2}{dt_i\varepsilon_i}\left(1-\sqrt{1-\frac{(c-i)^2}{c}}\right)+\frac{1}{\varepsilon_e^d}\frac{dmc^2}{dt_i\varepsilon_i}\left(1-\sqrt{1-\frac{(c-i)^2}{c}}\right)=$$
$$=\frac{\varepsilon_e^i+\varepsilon_e^d}{\varepsilon_e^i\varepsilon_e^d}\frac{dmc^2}{dt_i\varepsilon_i}\left(1-\sqrt{1-\frac{(c-i)^2}{c}}\right) \qquad \text{5E1}$$

ε_e^i and ε_e^d - are the intensity coefficients of the isotope and the donor;

ε_i - is the standard, time related intensity of the electron process.

If the electron process of the isotope is of less intensity than normal, isotope needs external electron process *blue shift* support; if more it gives off the surplus.

5E1 is with added electron process intensity, necessary for the element for having full collapse at the inflection point. The only disadvantage of this process is that this *blue shift* intensity will be missing later from the elementary process of the donor element.

If donors have been composed from number of elements, the quantum communication of elements and the rehabilitation of the isotope go in parallel. Key word is *equilibrium*!

The more is the number of components the higher is the chance to have close to equilibrium stage composition.

With simultaneous communication going on, the rehabilitation is

$$\frac{1}{\varepsilon_e^i}\frac{dmc^2}{dt_i\varepsilon_i}\left(1-\sqrt{1-\frac{(c-i)^2}{c}}\right)+$$
$$\frac{1}{\varepsilon_e^{d1}}\frac{dmc^2}{dt_i\varepsilon_i}\left(1-\sqrt{1-\frac{(c-i)^2}{c}}\right)+\frac{1}{\varepsilon_e^{d2}}\frac{dmc^2}{dt_i\varepsilon_i}\left(1-\sqrt{1-\frac{(c-i)^2}{c}}\right)= \qquad \text{5E2}$$
$$=\frac{\varepsilon_e^i\varepsilon_e^{d1}+\varepsilon_e^{d1}\varepsilon_e^{d2}+\varepsilon_e^i\varepsilon_e^{d2}}{\varepsilon_e^i\varepsilon_e^{d1}\varepsilon_e^{d2}}\frac{dmc^2}{dt_i\varepsilon_i}\left(1-\sqrt{1-\frac{(c-i)^2}{c}}\right)$$

In the case of no donors, the isotope is left alone to manage its intensity problems. It needs time and might be making damages to the environment.

With a single element as donor, the support of the isotope is given, but the donor element provides *blue shift* in line with its elementary process. The rehabilitation is controlled and speeded up to a certain level.

In the case the rehabilitation is supported by a mix with number of elements, the rehabilitation is more efficient: Rehabilitation with electron processes of many elements, acting in parallel.

The reason of the appearance/creation of an isotope is always complex:
The proton process cover of the neutron process can be damaged. The damaged proton process at the same time might be consequence of damaged neutron collapse. Reasons and consequences are in complex relation, depending on each other.

If the neutron process is driven by standard for the element electron process intensity and covered by its standard proton process intensity; in other words the intensity of the driven neutron process at the inflexion point corresponds to the standard value of the element – the intensity of the proton process (leaving the inflexion point) will be corresponding to the standards of this element. The intensities of the proton, electron and neutron processes represent the standard element.

<table>
<tr><td>S.
5.5</td><td></td></tr>
</table>

5.5

Stable matrix with increased intensity of the inflexion point

Two elements, one with proton process intensity dominance and electron process *blue shift* surplus and the other with intensive neutron process dominance cannot create strong structure of a mix for long even if the proportions would allow it.

Here below is the intensity formula of this kind of mix.

$$X \frac{dmc^2}{dt_p^A}\left(1-\sqrt{1-\frac{v^2}{c^2}}\right) + Y \frac{dmc^2}{dt_p^B}\left(1-\sqrt{1-\frac{v^2}{c^2}}\right) \neq$$

5F1

$$\neq X \frac{dmc^2}{dt_n^A}\sqrt{1-\frac{(c-i)^2}{c^2}}\left(1-\sqrt{1-\frac{v^2}{c^2}}\right) + Y \frac{dmc^2}{dt_n^B}\sqrt{1-\frac{(c-i)^2}{c^2}}\left(1-\sqrt{1-\frac{v^2}{c^2}}\right)$$

X portion of Element A provides electron process *blue shift* drive and proton process cover.
Element B provides electron and proton processes in portion Y.
Neutron processes are passive components, the elementary communication therefore will result in elementary relation in line with the acting electron process *blue shift* drives and the proton process covers of the communicating elements. Element A has more electron processes but of less intensity, Element B has less electron processes but of more intensity.

The resulting mix will be with electron process *blue shift* conflict, because of the surplus from Element A, but in proportions it will be less significant than the surplus alone within Element A.

Element B within the mix will be with less strength of its original solid structure, as the overall *blue shift* surplus will be softening it.

If a chain of infinite number of elements $(n-1)$ close to the stage of proton-neutron process equilibrium (above and below) is added to the mix, the characteristics of the mixture will significantly be improved. The integrated electron process intensity of this kind of mix can be written as:

$$\frac{1}{n-1}\frac{\varepsilon_2\varepsilon_{n-1}+\varepsilon_1\varepsilon_2+...+\varepsilon_1\varepsilon_{n-1}}{\varepsilon_1\varepsilon_2...\varepsilon_{n-1}}\frac{dmc^2}{dt_i}\left(1-\sqrt{1-\frac{(c-i)^2}{c^2}}\right)=$$

$$=\frac{dmc^2}{dt_i\varepsilon_e^{sum}}\left(1-\sqrt{1-\frac{(c-i)^2}{c^2}}\right) \qquad \text{5F2}$$

(The summarised intensities of the electron processes shall be divided by the number of electron impacts, since the integrated average intensity value shall relate to a single unit dm, addressing the intensity of the expansion of the mass of the electron process.) This intensity value equals to the intensity of the *blue shift* (radial acceleration of the expansion) at constant dt_i time system.

$$\varepsilon_e^{int}=\frac{\varepsilon_p}{\varepsilon_n}\sqrt{1-\frac{(c-i)^2}{c^2}}=\varepsilon_e^{sum} \qquad \text{5F3}$$

As result of the communication of the number of elements, the integrated electron process intensity coefficient will be quasi constant value ($\varepsilon_e^{int}\approx1\approx const$), meaning: it will be not sensitive to external *blue shift* impact.

Blue shift impacts from the *Quantum Membrane* will not be changing the relation of the proton and neutron processes. Therefore there will be no change in the structure of the elementary composition. It stays constant. In other words, the composition may be impacted by electron process *blue shift* without any elementary change, just increasing the electron *blue shift* conflict of the mix.

If ε_e^{int} the integrated intensity coefficient is not quasi constant and stable, the intensities of proton and neutron processes may be impacted. Any external electron process impact may initiate modification and restructuring of the composition, results in structural change.

If the average integrated intensity coefficient is constant – which actually means that the mass change (and *blue shift* impact) of the electron process is constant – the *work* of internal elementary processes might be of more or less value, addressing electron process impacts, but the relation of the proton and neutron process intensities of the mix will be not changing. Most importantly, the intensity of the inflexion point will be constant.

It might be formulated another way:
The intensity of the inflexion point is so stable that external impact (obviously within a certain range) cannot change it.

Any *external* impact to the stabilised structure means:

5F4
$$\frac{dmc^2}{dt_i \varepsilon_e^x}\left(1-\sqrt{1-\frac{(c-i)^2}{c^2}}\right) + \frac{1}{n-1}\frac{\varepsilon_2\varepsilon_{n-1}+\varepsilon_1\varepsilon_2+...+\varepsilon_1\varepsilon_{n-1}}{\varepsilon_1\varepsilon_2...\varepsilon_{n-1}}\frac{dmc^2}{dt_i}\left(1-\sqrt{1-\frac{(c-i)^2}{c^2}}\right)=w$$

The integrated intensity value of the electron process will be quasi not changing, but the work is increased. The increased electron process work results in increased value of elementary processes.

5F5
$$\frac{dmc^2}{dt_p \varepsilon_p}\left(1-\sqrt{1-\frac{v^2}{c^2}}\right) = \frac{dmc^2}{dt_n \varepsilon_n}\xi\sqrt{1-\frac{(c-i)^2}{c^2}}\left(1-\sqrt{1-\frac{v^2}{c^2}}\right)$$

Elementary processes happen in time and in balanced communication with the *Quantum System of Reference* and the *Quantum Membrane* (space).

Increased elementary communication as result of external *blue shift* impact increases the intensities of the proton and neutron processes, with still quasi constant integrated internal electron process intensity coefficient of the mix.

The intensity of the inflexion point is increasing but the relation of the intensities of the neutron and proton processes is without change. The intensities of the elementary processes of the mix have been increased without any structural change. The time frames of the neutron and proton processes are shortened. It means the energy capability of the mix is increased.

The *Quantum Membrane* with *blue shift* impact and the mix with the elementary process are in balance. It actually helps to manage the membrane function of the *Quantum System of Reference*. The intensity increase of the inflexion point is accumulating *blue shift* impact.

The increased intensity of the inflexion point means the equal increase of the intensities of the neutron collapse and the proton expansion – as the proton/neutron relation of the element or the created mix is not changing – still resulting in the same element, or the same composition of elements.

As the intensity increase of the mix is originating from the *Quantum Membrane*, as from external impact, the *Quantum Membrane* will be establishing a local, certain intensity reference of *quantum communication*, with increased *quantum communication* speed c, and the corresponding to $i = \lim a\Delta t = c$ electron process time system.

This way the intensity of the *inflexion point* will be corresponding to the parameters of this local *Quantum Membrane*.

In the case the external *Quantum Membrane* will be of less intensity, the mix, with the accumulated *blue shift* reserve of the local *Quantum Membrane* (a kind of energy) will be impacting it.

The explanation is that the intensity balance of the inflexion point does not need any more this high intensity *blue shift* impact. The return impact from the mix to the external *Quantum Membrane* can be of different formats including *blue shift* impact of certain frequency, addressing a certain *blue shift* request from aside or provide magnetic impact against magnetic features.

Releasing electron process *blue shift* impact with conflict from the mix, (through specific wire) may result in lighting effect: In the case of light, the *blue shift* released from the stable elementary structure via elementary communication of a wire might be in conflict at some points of the wire. This conflict may cause light effect.

> [Electron flow (electricity) is not other than moving *blue shift* impact. All stable elements lead electricity, as they do not need *blue shift*. All those with *Hydrocarbons*, with *Hydrogen* of not finished neutron process however do not.]

In the case of external intensity increase and mixtures with not close to equilibrium status, the result is either

- increased *blue shift* surplus – as proton process dominant mixtures will generate more *blue shift* surplus: The more intensified the neutron collapse (and with that the inflexion point) is the more is the electron *blue shift* impact; or
- without real impact, as while the increased *blue shift* impact would drive the neutrons of the neutron process dominant mixture, but without proper proton process cover this could not happen.

5.6
Electron flow in space and time

S. 5.6

What we call "electron flow" is propagating electron *blue shift* impact. Constant proton process and electron process *blue shift* surplus generates impact to the environment as "electron flow". Electron processes however do not move; they belong to certain *Strong Interrelations*. Their impact is the one propagating through the *Quantum System of Reference* in any direction.

Since proton and neutron processes with different intensities communicate without any problem in infinite environment, "loosing" *blue shift* impact from the communication does not mean missing components of the drive of the actual process. The infinite number of components always ensures the continuity of the process. Space and different time systems guarantee continuity.

The *blue shift* surplus as impact is propagating in the *Quantum System of Reference*. There are significant good tools for supporting this communication, like those elements with slight neutron process dominance.

There are also those, with wrong electricity flow function, like those with heavy neutron process dominance, where all provided electron *blue shift* is "swallowed" by the element as drive to the neutron process.

The main point here is about the flow of the impact of the *blue shift,* rather than the electrons themselves. *Blue shift* drive makes neutrons to collapse *only if proton process cover is also available*. Without proton process, the *blue shift* impact increases the intensity of the *quantum communication*, as it loads the *Quantum Membrane*.

In the case of increased intensity of the *Quantum Membrane* there could be couple of options.
-    the impact is flowing through without any real effect;
-    the intensive electron process *blue shift* impact increases the intensity of the *Quantum Membrane* and communication, which might cause damages in the elementary structure.

Electron process *blue shift* impact flowing within wires means increased membrane function. Increased membrane function results in increased proton process and neutron process intensities.
Increased neutron process intensity needs increased proton cover. The increased intensity of the processes needs sufficient quantum space or/and time – electron flow.

6
Quarks
the measured indications of the proton, neutron and electron processes

S.6

Proton, neutron and electron processes represent continuous flow of changing *quark* statuses. With reference to Section 25.2 of Book 3 on *Quantum Engine* we measure certain periods of this process and mark it as *top*, *charm*, *up*, *bottom*, *strange* and *down processes*. Proton process, as acceleration (transformation of mass into energy) and neutron process, as collapse (re-transformation of the energy status into mass) driven by the electron process are in energy and mass balance.

Ref.
Book
3
S.25.2

The expansion-collapse balance and chain is shown on Diagram 6.1.

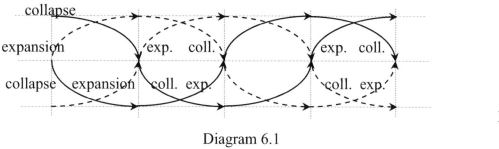

Diagram 6.1

Diag.
6.1

The *proton* process is built up from *two* (t-c-u) *acceleration* (expansion) processes and a *single* (d-s-b) process, which is *collapse*. The *neutron* process in the contrary has *two* (d-s-b) and a *single* (t-c-u) processes.

Diagram 6.2 demonstrates the particles as processes. Their conventionally measured mass value is equivalent to the measured impact of the process.

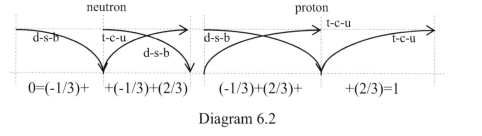

Diagram 6.2

Diag.
6.2

The numbers in the diagram illustrate the charge of the "particles".
In fact these mean: proton provides the energy during the expansion (as *1*) and neutron takes and uses it by its collapse (and it becomes neutral: *0*).

Electric charges of the chain of quark processes are as follows
Down: –1/3e; *Strange*: –1/3e; *Bottom*: –1/3e
Top: +2/3e; *Charm*: +2/3e; *Up*:   +2/3e

The other version of the relations is shown in Diagram 6.3:

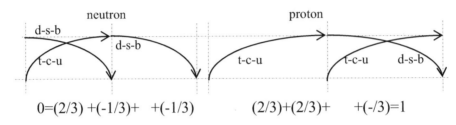

$$0 = (2/3) +(-1/3)+ \ +(-1/3) \qquad (2/3)+(2/3)+ \ \ +(-/3)=1$$

Diag.
6.3

Diagram 6.3

The concern with Diagrams 6.2 and 6.3 is the *inflexion point,* the fluent transfer from one *hadron* process to the other. Collapse continues with expansion and expansion with collapse. The transition between the two depends on the intensity of the processes.

Real inflexion happens between processes and anti-processes as *sine* and *cosine* do:

> neutron collapse continues with anti-proton expansion; and
> anti-neutron collapse continues with proton expansion.

Each proton process (*normal* and *anti*) of the chain is developing into electron process. (There is no difference in the electron processes.)
Diagram 6.4 shows all those options of the *quark*-processes, which have been found as components of elementary particles.

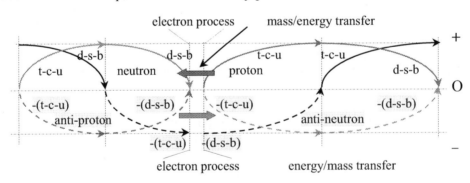

Diag.
6.4

Diag.6.4

The opposite direction of processes is equivalent to anti-processes:
$$(u-c-t) = -(t-c-u); \quad (b-s-d) = -(d-s-b)$$

(t-c-u) means: (d-s-b) means:
    the expansion is positive   the collapse is positive
(u-c-t) means: (b-s-d) means:
    the expansion is negative   the collapse is negative

The opposite direction means, that the charges of quark processes are also taken with opposite signs.

Processes relating to the upper part of the diagram 6.4:

Neutron process   Proton process:

$$0 = \left(-\frac{1}{3}\right) + \left(-\frac{1}{3}\right) + \left(\frac{2}{3}\right)$$ $$\left(\frac{2}{3}\right) + \left(-\frac{1}{3}\right) + \left(\frac{2}{3}\right) = 1$$ 6A1

Processes relating to the lower part of the diagram 6.4:

Anti-proton process   Anti-neutron process:

$$-1 = \left(-\left(-\frac{1}{3}\right)\right) + \left(-\frac{2}{3}\right) + \left(-\frac{2}{3}\right)$$ $$\left(-\left(-\frac{1}{3}\right)\right) + \left(-\left(-\frac{1}{3}\right)\right) + \left(-\frac{2}{3}\right) = 0$$ 6A2

The existence of processes and anti-processes in the nature is explained by the direction and the quality of mass/energy and energy/mass transfers.

The mass/energy and/or energy/mass content of elementary processes are constant (with the obvious difference of the developing in each cycle *entropies*).

The *Strong Interrelation* however

➢ between proton and neutron processes means:
Increasing *mass*/energy content of the neutron for the count of the decreasing *energy*/mass content of the proton.
Proton is losing on *energy* and neutron is gathering on *mass*. The summarised mass/energy content of the relation remains unchanged.

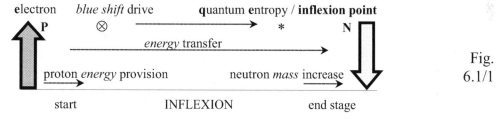

Fig. 6.1/1

➢ between anti-proton and anti-neutron processes means:
Increasing *energy*/mass content of the anti-neutron for the count of the decreasing *mass*/energy content of the anti-proton.
Anti-proton losing on *mass* and anti-neutron is gathering on *energy*. The summarised mass/energy content of the relation remains unchanged.

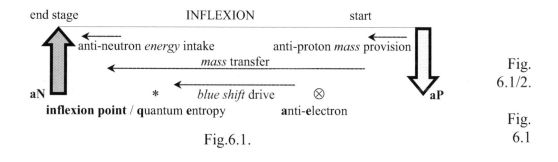

Fig. 6.1/2.

Fig. 6.1

Fig.6.1.

6A3 $E = mc^2$ characterises the full energy/mass or mass/energy content, as

6A4 $e = \dfrac{dmc^2}{dt}$ characterises also the both, the summarised intensities of the *mass* and *energy* transformation.

From the point of view of the balance, mass and energy values are identical. But as quality they are different!

The *Quantum System of Reference* is composed from *quantum*, the unique and common appearance of the smallest as possible content of energy/mass or mass/energy values, corresponding to the entropy value of the cycle.

Quantum Membrane is impacted by electron process *blue shift* and the continuity law of any mass-energy (or energy-mass) exchange guaranties, there is no loss in the summarised *Einstein's* formula: $E = mc^2$

The communication within the *Quantum Membrane* is mass/energy or energy/mass based, as the *quantum* itself represents both mass and energy. The definition of *quantum* cannot only be connected to energy!

Energy-quantum as such dos not exists. *Quantum* is mass and energy or energy and mass. There is no priority among the two, as there is no priority between time and space as well.

Ref.
Fig.
6.1/1

Quantum Membrane transfers the impact and the mass increase of the neutron process is compensated by the energy release of the proton process through the quantum space: With reference to Fig.6.1/1, the energy of the proton is incorporating into the neutron as mass.

As result,

the neutron as recipient is of full mass/energy or energy/mass, at the same time the other donor end with proton and electron processes is at the stage of entropy.

If receiving energy/mass from a certain source, which is incorporating as mass/energy into another was a process, called "positive"; the release of this mass/energy with its transfer back (as the balance and the continuity of the process need it) should be with "negative" prefix for sure. But provision of mass/energy is proton function, to be driven by *blue shift* drive and neutron demand. With reference to Fig.6.1/2 this is the *anti-process* of the elementary relations, with all quark processes with negative signs.

Ref.
Fig.
6.1/2

The same way and logic, the anti-neutron process, reaching the inflexion point continues as proton process.

In "process – anti-process" relations the balance, the direction of events and the inflexions clearly specify the case.

The energy/mass transfer of the "normal" and the mass/energy transfer of the "anti" processes as subjects and constant responses to each other do not add anything principal to the explanation. They might be used in either ways as distinguishing characteristics.

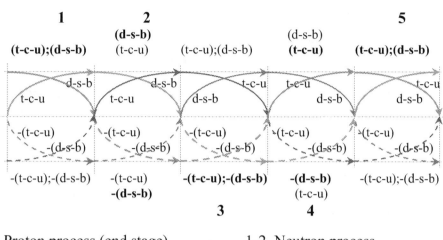

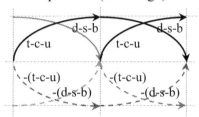

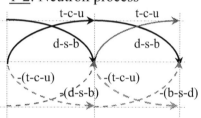

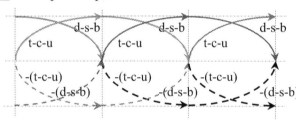

1. Proton process (end stage)   1-2. Neutron process

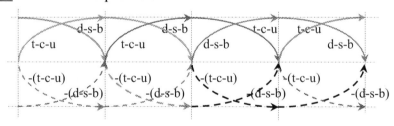

Diag. 6.5/1

Diag. 6.5/2

2-3. Anti-proton process:

Diag. 6.5/3

3-4. Anti-neutron process:

Diag. 6.5/4

4-5. Proton process:

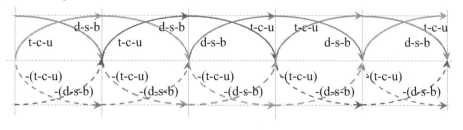

Diag. 6.5/5

Diag. 6.5

Diag. 6.5

Diagram 6.5 demonstrates a full cycle of the process from proton process to proton process. Each colour represents the continuity of the *down-strange-bottom-top-charm-up* chain of quark processes. The lines in dash are for the anti-processes.

Each proton and anti-proton processes end up with electron processes. In both cases the electron process is about generation of *blue shift*. Therefore there is no difference. Electron processes are the drives of the elementary communication. (Anti-electron process would mean generation of *red shift*, which is here not the case.)

The end of *each* neutron process is the start of an *anti-proton* process. And *each anti-neutron* processes is followed by proton process.

This means that neutron process has its sufficient load from the *Strong Interrelation*, driven by electron processes. The accumulated by the collapse mass/energy is expanding during the followed proton and anti-proton process.

Electron processes at both, "normal" and "anti" ends end up in *quantum entropy* status which continues as neutron or anti-neutron.

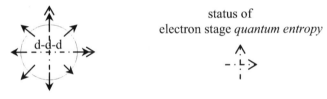

status of
electron stage *quantum entropy*

Diag.
6.6   Diagram 6.6

Diagram 6.4 initiates the process and the energy transfer:
Transformation of the energy/mass of the proton extension into the mass/energy of the neutron collapse; and the transformation of the mass/energy of the anti-proton expansion into the energy/mass of the anti-neutron collapse.

There is no difference in absolute terms between the proton and the anti-proton processes, with taking into account the entropy is eliminated:

6A5
$$\left| \frac{dmc^2}{dt} \left(1 - \sqrt{1 - \frac{v^2}{c^2}} \right) \right|; \quad \text{and} \quad \left| -\frac{dmc^2}{dt} \left(1 - \sqrt{1 - \frac{v^2}{c^2}} \right) \right|$$

> *Negative mass gradients* in practice usually mean the increase of energy or work load – in global terms the utilisation of mass.
> *Negative energy gradients* as such have not been connected with mass increase, while the symmetry would need it and it makes sense.

With reference to Diagrams above, for the better presentation of the case the neutron and proton (and also the anti-proton and anti-neutron) processes are taken of quasi equal intensity with the electron processes at both sites have been left out.

The relation between processes and "anti-processes" gives the explanation to the understanding of particles and anti-particles. On particle based approach this category has never been really clarified. Process basis specifies the relation and addresses the gradient of the intensity change as responsible for the definition.

The *inflexion point* of the process is at "**O**".
"+" means normal processes
"–" means the anti-processes.

6.1

Symmetry and intensities of the processes

S.

6.1

Proton and neutron (and also anti-proton and anti-neutron) processes have been introduced in the diagrams as of being in full symmetry.
This is however not the case. They are with certain shift in intensities to each other, which are the specifics of each element. The symmetry is only used for easy interpretation.

Intensities are changing constantly even during the same process. This is the reason and explanation of finding quarks with different frequencies.
Quark processes with low intensity and long life time have been found more often (as *up* in proton process and *down* in neutron process) than others with more intensity and shorter process duration (as *charm* and *strange* or *top* and *bottom*).

The absolute balance in proton and neutron processes is guaranteed at any time point:

$$\frac{dmc^2}{dt_o \varepsilon_p}\sqrt{1-\frac{v^2}{c^2}}\left(1-\sqrt{1-\frac{v^2}{c^2}}\right) = \frac{dmc^2}{dt_o \varepsilon_n}\xi\sqrt{1-\frac{v^2}{c^2}}\sqrt{1-\frac{(c-i)^2}{c^2}}\left(\sqrt{1-\frac{v^2}{c^2}}-1\right) \qquad \text{6B1}$$

But the intensities of the proton and neutron processes are different:

$$\frac{dmc^2}{dt_o}\sqrt{1-\frac{v^2}{c^2}}\left(1-\sqrt{1-\frac{v^2}{c^2}}\right) \neq \frac{dmc^2}{dt_o}\sqrt{1-\frac{v^2}{c^2}}\sqrt{1-\frac{(c-i)^2}{c^2}}\left(\sqrt{1-\frac{v^2}{c^2}}-1\right) \qquad \text{6B2}$$

Proton intensity is: $e_p = \dfrac{dmc^2}{dt_p}\left(1-\sqrt{1-\dfrac{v^2}{c^2}}\right) = \dfrac{dmc^2}{dt_o}\sqrt{1-\dfrac{v^2}{c^2}}\left(1-\sqrt{1-\dfrac{v^2}{c^2}}\right)$ \qquad 6B3

Neutron intensity is: $e_n = \dfrac{dmc^2}{dt_n}\sqrt{1-\dfrac{(c-v)^2}{c^2}}\left(\sqrt{1-\dfrac{v^2}{c^2}}-1\right) =$

$$= \frac{dmc^2}{dt_o}\sqrt{1-\frac{v^2}{c^2}}\sqrt{1-\frac{(c-i)^2}{c^2}}\left(\sqrt{1-\frac{v^2}{c^2}}-1\right) \qquad \text{6B4}$$

The specific, element related intensity coefficient of the electron process is dimension less:

6B5
6B6

$$\varepsilon_x = \frac{\varepsilon_p}{\varepsilon_n}\sqrt{1-\frac{(c-i)^2}{c^2}} \quad \text{and} \quad E_e = \frac{dmc^2}{dt_i\varepsilon_i\varepsilon_x}\left(1-\sqrt{1-\frac{(c-i)^2}{c^2}}\right)$$

The intensities of the expansion and the collapse have their gradients of the change.

Ref.
S.5.3
Diag.
5.4

While the element related electron process intensity is constant and characteristic of the element, electron function has its variety in the values of acceleration, reference to Section 5.3 and Diagram 5.4, since: $(c-i) = a\Delta t$

The process can be measured and distinguished as: *electron* (e) *muon* (μ) and *tau* (τ). The function is the same just the measured timeframe and the value of the acceleration effect of the process are different!

Not just quarks have 6 different flavours, but the electron drive function can also be distinguished, based on the parameters of the process.

As direct consequence of the natural deviation of the full proton-electron-neutron balance there are neutron processes with proton cover demand, named after the basic electron function: *electron neutrino* (v_e), the *muon neutrino* (v_μ) and the *tau neutrino* (v_τ).

Neutrinos are not isotopes. They are "tools" of the natural correction.

With reference to Fig.4.2, the shift in intensities can be presented as:

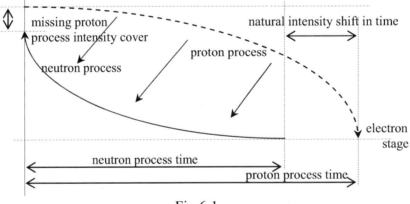

Fig.
6.1

Fig.6.1

Diagram 6.7 demonstrates an integrated cycle with all quark process options. It gives options to elements to be connected to other elements.

(The cycle is integrated indeed, but this is not the full scope of quark processes.)

Quark processes are taken as ideal and presented as symmetrical ones without escorted by electron functions of different type and balance deviations in form of neutrinos.

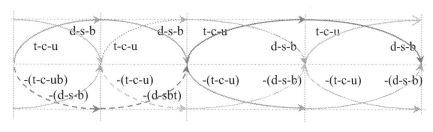

Diagram 6.7

Quark expansion process starts at the inflexion point, at the status of full collapse of the matter.

With reference to the full collapse from the balance equation follows that at this status the value of the *quantum entropy* is:

$$qe = \frac{dmc^2}{dt_o}\sqrt{1-\frac{(c-i)^2}{c^2}}$$

6C1

At this status and at this time system 6C1 is the intensity of the quantum entropy, the generator of the energy *quantum* in 6C2:

$$q = \frac{dmc^2}{dt_o}\left(1-\sqrt{1-\frac{(c-i)^2}{c^2}}\right)$$

6C2

At the other end, at full expanded status the collapse starts when the *blue shift* impact of the *electron* process insufficient any more for driving the collapse:

$$\frac{dmc^2}{dt_o}\sqrt{1-\frac{i^2}{c^2}}\left(1-\sqrt{1-\frac{(c-i)^2}{c^2}}\right) < q_e^i$$

6C3

The intensities in 6C2 and 6C3 represent equal absolute *quantum* values. As intensities they relate to different time systems:

$$dt_o = dt_i\sqrt{1-(i^2/c^2)}$$

6C4

Processes happen for the count of the internal energy and mass balance, the *Strong Interrelation* of matter.

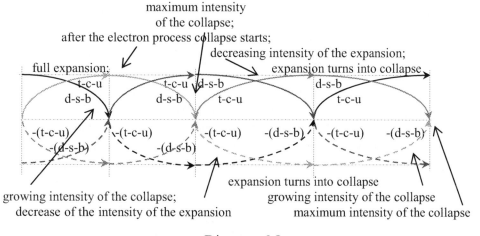

Diagram 6.8

6.2

Quantum Word

Quantum world is not about particles. Internal components of protons, neutrons and electrons are *quarks*. Quarks are detected as particles, but they are processes as well. They are measured as particles in our time system of reference. Types of *quarks* represent different intensities of the sphere symmetrical expanding acceleration and the accelerating collapse.

The most frequently measured quarks with low quantum impact [measured as mass (effect)] are those with low intensity of the process, like *up* and *down*.

Quarks with rare detection are of high(er) intensity of the quantum impact [measured as mass (effect)] and with high(er) intensity of the process – obviously with less detectable "period" of their process time, like *top* and *bottom* type quarks.

This finding well correlates with the process based definition of protons, neutrons and also of the *electron* functions – the *muon* and *tau* as well.

In proton (process) the sphere symmetrical expanding acceleration (*t-c-u*) is the dominant, as the other two processes [(t-c-u) and (d-s-b)] are in balance.

In neutron (process) the sphere symmetrical accelerating collapse *(d-s-b)* is the dominant, while the other two processes [(d-s-b) and (t-c-u)] are in balance, as they are in the proton as well.

In electron function the time system is constant: $i = \lim a\Delta t = c$, but the intensity of the process is impacted by proton/neutron intensity relations and is changing from *electron* to *tau* via *muon*.

6D1
$$\frac{dmc^2}{dt_i \varepsilon_i \varepsilon_x}\left(1 - \sqrt{1 - \frac{(c-i)^2}{c^2}}\right) = \frac{dn}{dt_i \varepsilon_i \varepsilon_x} q = f_x q$$

While the equation with the definition of the time(-system) (dt_i) and its intensity consequence (ε_i) gives absolute value, (ε_x) makes 6D1 unique since the function in 6D1 only characterises element *x*.

Up type quark process represents the end stage of the proton process, with speeding up its expansion process approaching $i = \lim a\Delta t = c$.

The cycle continues with electron function: *tau* the most intensive and less measurable, through *muon*, with rare measurement and finally *electron* of less intensity, but dominant existence with full expanded status of (*d-d-d*).

This is the start of the neutron process, with detected *u-d-d* status. *Down* quarks represent the starting stage of the neutron process with low intensity and long measurability in our time system.

Neutron process approaches the inflexion point and the anti-proton process starts. The anti-proton and anti-neutron processes through the electron process *blue shift* drive rearrange the internal mass/energy structure of the process and the proton process at the inflexion point starts with full energy.

Electron process in fact has only a one-way function: drive the neutron collapse as transforming into neutron process itself, reaching status: *d-d-d*.
There is no *anti-electron* function at the end of the anti-neutron process.
  Anti-electron impact as such should be identified as *red-shift*.
  The difference is in the detection: *blue shift* is the increase, *red shift* is the decrease of the impact. This way all *tau, muon* and *electron* processes have their *anti-process* components as well.

Within the proton, anti-proton and within the neutron and anti-neutron processes separately, the expansion and the collapse have been in internal balance. The balance is for the 2/3 part of the expansion and the collapse. The relation is of 1/3 to 1/3 in balance of both.

The remaining 1/3 part of the collapse and 1/3 part of the expansion should be in balance at the level of elementary communication, the *Strong Interrelation.*
The relation of the anti-proton and anti-neutron processes is the same, just as the gradients of intensities are the opposite to the directions of the energy transfer are also the opposite.

The electron process function within the element (in all its measured three forms) is *blue shift* impact, the *Weak Interrelation.* It has its drive impact to the collapse within proton, neutron, anti-proton and anti-neutron processes as well.

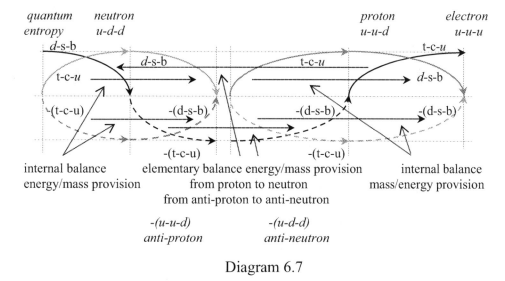

Diagram 6.7

Endless continuity can be demonstrated by the diagrams above.

The *blue shift* impact of the electron function – the impact of the expanding mass against the *Quantum System Reference* establishes a *Quantum Membrane* of certain intensity. This impact can also be called as *gluon*, but in fact this is the *Quantum Membrane* impacted by the electron process.

The energy/mass balance can be demonstrated in the following diagrams:

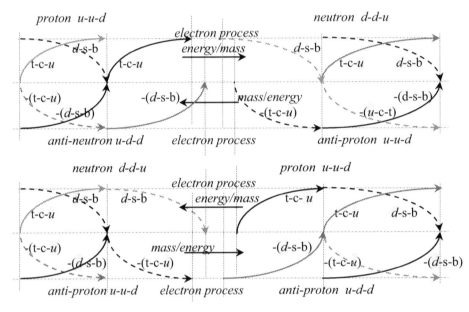

Diag.
6.8   Diagram 6.8

S.   **6.3**

6.3   **Mass and Energy Balance**

Mass and *energy* can be understood as measured statuses or appearances of matter in time. Proton (and anti-), neutron (and anti-) processes and the electron function represent the *time (flow)*, since events happen in time.

The intensities of the mass/energy expansion and the energy/mass collapse change in time. The status is changing and the intensities of the change are also changing.

6E1   The intensity of the change, initially $\dfrac{dmc^2}{dt}$, becomes modified

6E2   on its own expansion to another intensity value: $\dfrac{dmc^2}{dt}\sqrt{1-\dfrac{v^2}{c^2}}$.

The question is where the "missing" part of the change is disappearing?

6E3 $$\frac{dmc^2}{dt} = \frac{dmc^2}{dt}\left(1-\sqrt{1-\frac{v^2}{c^2}}\right) + \frac{dmc^2}{dt}\sqrt{1-\frac{v^2}{c^2}}$$

The first component at the right hand side of the balance in 6E3 is the transformed intensity (of the proton process), the other is the remaining one. We can call this transformed (missing) part, natural consequence of mass transformation: *energy*. There is no way this intensity (=energy) could be accumulated.

This "intensity-transformation" is permanent and the interrelation between proton-neutron processes – the transformation of mass (of the matter) into energy and its retransformation back into mass – ensures the balance.

The balance is: accelerating expansion against accelerating collapse.

The developing intensity (energy) of the *proton* process is incorporating into the *neutron* process (mass).

This is the *Strong Interrelation*, which ensures the continuity.

And the developing by the anti-proton process mass/energy intensity is incorporating into the anti-neutron process as energy/mass.

reaching

The gradient of the energy/mass change of the expansion is negative	$-\dfrac{dm}{dt}_{exp}$	$\lim \dfrac{dm}{dt}_{exp} = 0$	6F1
providing mass/energy to the collapse, which is positive:	$\dfrac{dmc^2}{dt}_{col}$	$\lim \dfrac{dmc^2}{dt}_{col} = \infty$	6F2

Expansion starts with the intensity at the collapse	$\lim \dfrac{dmc^2}{dt}_{col} = \infty \quad = \dfrac{dm}{dt}_{exp}$	6F3
Collapse starts with the intensity at the expansion	$\lim \dfrac{dm}{dt}_{exp} = 0 \quad = \dfrac{dm}{dt}_{col}$	6F4

The intensity of the electron process drive is changing from *tau* to *electron* via *muon*, in line with the decreasing value of acceleration and the increasing time component, while in principle the time system is constant.

$$e_x = \frac{dmc^2}{dt_i \varepsilon_x}\left(1 - \sqrt{1 - \frac{(a\Delta t)^2}{c^2}}\right) \qquad \text{6F5}$$

The *blue shift* drive demand is established by the intensity coefficient of the electron process: $\qquad \varepsilon_x = \dfrac{\varepsilon_p}{\varepsilon_n}\sqrt{1 - \dfrac{(c-i)^2}{c^2}} \qquad$ 6F6

The beauty of this balance is that the *Quantum System of Reference* is result of the accumulating natural non-balance of the *Strong Interrelation*.

The components of the transformation cannot turn in full into the other.

Ref
6C2 This is the *entropy* rule – the rule of any real balance relation. The developing energy *quantum*, reference to 6C2 consequence of this natural non-balance of the processes composes the *Quantum System of Reference.*

If the intensities of the energy/mass generation (proton, anti-proton) and the mass/energy utilisation (neutron and anti-neutron) are different, processes happen for different time periods.

The drive of the *Strong Interrelation* is the process of the electron function:

It initiates the collapse. The more electrons have been generating, the more is the drive surplus of the collapse.

Electron function as process is of constant time system at constant speed value of $i = \lim a\Delta t = c$. It seems as it is also of constant intensity.

It is however not the case!

The electron function as process is *blue shift* impact to the *Quantum System of Reference* – creating with that the *Quantum Membrane* – of constant time system.

[Proton and neutron processes (and anti-proton and anti-neutron processes as well) are transformations with different transformation gradients and *without* impacting the *Quantum Membrane*.]

Electron function (including *muon* and *tau* as well) is sphere symmetrical expanding accelerations at constant speed. Different values of acceleration impact the *Quantum System of Reference* for different time periods. More value of acceleration is acting for less time.

The electron function as process is acting within constant time system, with variety of acceleration values for corresponding time durations of the impact to the *Quantum Membrane*.

This way, while the intensity of the electron process is constant and characteristic of the element, the acting format of the impact to the *Quantum*

Ref
6B6 *Membrane* varies.

With reference to 6B6 the intensity of the electron process means the *element* itself.

Beside its intensity feature, electron function has its other characteristic as well. This is about electron process *surplus* or "deficit".

The *Strong Interrelation* of elements controls the case:

➢ If the expansion process is more intensive, there will be always electron process surplus, since its utilisation remains behind its generation.

➢ If the collapse process is more intensive, the electron process will be acting always in deficit conditions.

Elements themselves are stable and are in internal balance status either with electron surplus or electron deficit.

As distinguishing characteristic however the value of the surplus and/or the deficit determines the main "chemical" capabilities of the element.

With the expansion being more intensive than the collapse the element is with electron process and *blue shift* surplus. Elements with neutron and anti-neutron process intensity dominance have in the contrary *blue shift* "deficit". Elements with dominant neutron and anti-neutron process intensity and electron *blue shift* deficit will be utilising available *blue shift* surplus.

Blue shift communication amongst elements improves the balance. Full balance may create unbreakable elementary structure, with no communication with the external environment.

"Chemical" characteristics of elements depend on the intensity balance of the proton-neutron, anti-proton–anti-neutron processes and the surplus or the deficit of the electron *blue shift*. The keys are

the intensity of the transformation of mass (expansion):	and the intensity of the re-transformation of energy into mass (collapse):	
$$\frac{dmc^2}{dt_o}\left(1-\sqrt{1-\frac{v^2}{c^2}}\right)$$		6G1
	$$\frac{dmc^2}{dt_o}\sqrt{1-\frac{i^2}{c^2}}\sqrt{1-\frac{(c-i)^2}{c^2}}\left(\frac{1}{\sqrt{1-\frac{v^2}{c^2}}}-1\right)$$	6G2

$$\frac{dm}{dt_o}\sqrt{1-\frac{v^2}{c^2}}$$ is the intensity of the mass transformation in time; 6G3

$$\frac{dm}{dt_o}c\sqrt{1-\frac{v^2}{c^2}}$$ is the intensity of the momentum of the mass transformation, propagating within the *Quantum System of Reference* without impacting it. 6G4

c is the speed of quantum communication of the *Quantum Membrane.*

Why the intensity of the momentum of the energy/mass-mass/energy transformation, the *Strong Interrelation* does not impact the *Quantum Membrane*, while electron function *blue shift* does?

The *Strong Interrelation* is the connection between proton and neutron (anti-proton and anti-neutron) processes, parts of the same balance.

Energy and mass in normal circumstances cannot be accumulating separately. Only in the case of damaged balance.

Normal and balanced collapse, driven by electron *blue shift* impact is only possible if the incorporating mass/energy exactly corresponds to the energy/mass released by the expansion:

$$\left|\frac{dmc^2}{dt_p\varepsilon_p}\left(1-\sqrt{1-\frac{i^2}{c^2}}\right)\right|=\left|\frac{dmc^2}{dt_n\varepsilon_n}\xi\sqrt{1-\frac{(c-i)^2}{c^2}}\left(\sqrt{1-\frac{i^2}{c^2}}-1\right)\right|$$ 6G5

The entropy part (6C1) is establishing the *energy quantum* (6C2)!

S.
6.4

6.4
The beauty of the quantum approach
without words

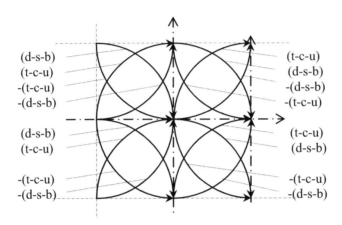

Diag.
6.9

Diagram 6.9

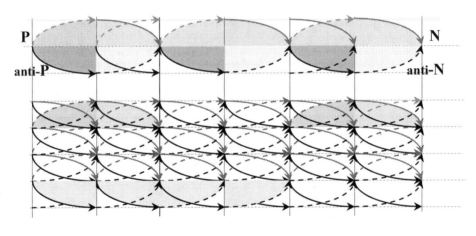

Diag.
6.10

Diagram 6.10

7
Magnetic Features

Carbon steel or normal iron rod is taken with wires winded all around the cylindrical surface. In the case of "electron flow" within the wire, the rod has magnetic features. Let us see, what is the explanation of the case, using *quantum energy and mass balance* approach?

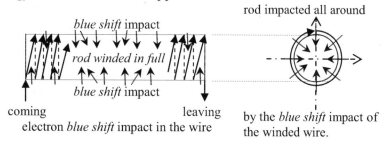

Fig.7.1

Fig. 7.1

Electricity flow is moving *blue shift* impact. The incoming and propagating *blue shift* impact intensifies the *Quantum Membrane* of the elementary processes within the wire. The intensified *blue shift* impact of the elementary processes of the wire through its external surface is also impacting the *Quantum Membrane* of the winded rod in any possible direction all around the rod.

The more is the number of the winds of the coil, the more is the impact.

The *Quantum Membrane* within the internal space of the rod is under increased external *blue shift* impact.

The summarised external *blue shift* impact can be written as:

$$N \frac{dmc^2}{dt_i \varepsilon_x} \left(1 - \sqrt{1 - \frac{(c-i)^2}{c^2}}\right) = N \frac{n}{dt_i \varepsilon_x} q$$	N is the number of electron *blue shifts*, impacting the *Quantum Membrane* of the rod

7A1

The *Quantum System of Reference* of the rod is of infinite number of elementary processes. If a single impact within the *quantum space* of the rod with n quantum, would result in certain quantum impact intensity, N *blue shift* impact would result in a proportional to N, $f(N)$ – function of N – times more intensity value. 7A1 can be written in its absolute form as:

$$\frac{dmc_{f(N)}^2}{\frac{dt_i}{f(N)} f(N) \varepsilon_i \varepsilon_x} \left(1 - \sqrt{1 - \frac{(c_{f(N)} - i_{f(N)})^2}{c_{f(N)}^2}}\right) = \frac{dmc_s^2}{dt_{is} \varepsilon_{is} \varepsilon_x} \left(1 - \sqrt{1 - \frac{(c_s - i_s)^2}{c_s^2}}\right)$$

7A2

ε_x is the electron process intensity coefficient. This value as a function is not changing, as the elementary characteristics of the rod remain unchanged.

$f(N)\varepsilon_i$ is intensity value under external impact, related to the time flow.

The intensity increase of the *Quantum Membrane* means, its internal, *quantum-to-quantum* communication is more intensive: its absolute quantum impact is of more value. The load of the *Quantum Membrane* is increased within the rod.

➤ If taken that $f(N) > 1$, the electron process time system of the winded road is speeded up, meaning: shorter measured time. The increased speed c of quantum communication results in increased speed of electron process, as $i = \lim a\Delta t = \lim c$.

➤ If taken that $f(N) < 1$, the time system of the winded road is slowed down, the time count is longer. The electron process this way is of less intensity, less speed of quantum communication and longer time system of the electron process.

As consequence of the increased electron flow and acting *blue shift* impact within the coil, the *absolute* value of the *blue shift* load of the *Quantum Membrane* of the winded rod relative to its normal status is increased.

In line with this the absolute values at the two sides of 7A3 are different.

7A3
$$\frac{dmc_s^2}{dt_{is}\varepsilon_{is}\varepsilon_x}\left(1 - \sqrt{1 - \frac{(c_s - i_s)^2}{c_s^2}}\right) > \frac{dmc^2}{dt_i\varepsilon_i\varepsilon_x}\left(1 - \sqrt{1 - \frac{(c - i)^2}{c^2}}\right)$$

At the same time however: $dt_{is}\varepsilon_{is}\varepsilon_x = dt_i\varepsilon_i\varepsilon_x = \varepsilon_x$ as $dt_{is}\varepsilon_{is} = dt_i\varepsilon_i = 1$,

meaning: the load – as expected – does not have effect at the elementary proton/neutron process intensity relation of the rod. $\lim\{(c - i)/c\} \cong \lim\{(c_s - i_s)/c_s\} = 0$ taken with no significant impact;

at the results; and *dm* is comparative value, taken equal in both cases.

There is only one way for managing the difference: c, the speed of *quantum communication* of the *Quantum Membrane* (within the internal structure of the rod) is of higher value.

7A4 $c_s > c$ the *quantum communication* in the loaded *Quantum Membrane* of the rod of the magnet is more intensive.

As of the time system concern of the winded road above, this statement needs more explanation:

7A5
7A6
$$dt_i = \frac{dt_o}{\sqrt{1 - \frac{i^2}{c^2}}}; \quad ds_i = \frac{ds_o}{\sqrt{1 - \frac{i^2}{c^2}}}$$

In the case of speed values $i = \lim a\Delta t = c$ the time and space effects are *negligible* and certainly behind the increase of c and i.

The *time system* within the road represents increased speed value of

7A7   *quantum communication*: $i_s = \lim a\Delta t = c_s$ and $\lim(c_s - i_s) = \lim(c - i) = 0$

The speed of quantum communication within the winded road is increased, as the *blue shift* load of its *Quantum Membrane* has also been increased!

More intensive quantum communication means the number of quantum *communications* within the road has been increased: $n_{is} > n_i$

$$\frac{dmc_s^2}{dt_{is}\varepsilon_{is}\varepsilon_x}\left(1-\sqrt{1-\frac{(c_s-i_s)^2}{c_s^2}}\right)=\frac{dn_{is}}{dt_{is}\varepsilon_{is}\varepsilon_x}q \qquad \text{7B1}$$

The *intensity* of the electron process of the rod in normal circumstances is:

$$e_i=\frac{dmc^2}{dt_i\varepsilon_x}\left(1-\sqrt{1-\frac{(c-i)^2}{c^2}}\right)=\frac{dn_i}{dt_i\varepsilon_x}q \qquad \text{(rod without the coil)} \qquad \text{7B2}$$

The *intensity* of the electron process in intensified and loaded status of the *Quantum Membrane* of the rod (with winded coil) is:

$$e_{is}=\frac{dmc_s^2}{dt_{is}\varepsilon_x}\left(1-\sqrt{1-\frac{(c_s-i_s)^2}{c_s^2}}\right)=\frac{dn_{is}}{dt_{is}\varepsilon_x}q \qquad \text{7B3}$$

$$c_s>c;\ dt_{is}\approx dt_i;\ \lim(c-i)=\lim(c_s-i_s)=0 \ \text{ as } \ i_s=\lim c_s \text{ and } i=\lim c$$

$$\text{and} \qquad e_{is}>e_i \qquad\qquad \text{7B4}$$

c and c_s are the velocities of *quantum communication*, the propagation of
  the *quantum impact*, expressed in relation to the *time* system of the event.
q *(quantum)* is result of quantum entropy, the remaining, accumulating
  mass/energy balancing part of the infinite chain of elementary cycles,
  establishing the *Quantum System of Reference.*

The increase of the speed of the quantum communication from c to c_s is the key. Increased *Quantum communication* covers more "*quantum space*" for *unit* period of time.
Quantum Membrane with increased intensity has increased workload and "energy". More intensive *Quantum Membrane* means more quantum impact – quantum communication – for unit period of time.

Permanent external *blue shift* impact keeps the *Quantum Membrane* loaded: Proton and neutron processes within the impacted rod – as consequence of the increased quantum communication – are of increased intensity; more intensive than in normal circumstances:
The intensity of the *proton process* at the inflexion point:

Ref.
S.8
S.9

$$\frac{dmc_s^2}{dt_p^s}\left(1-\sqrt{1-\frac{i_s^2}{c_s^2}}\right)>\frac{dmc^2}{dt_p}\left(1-\sqrt{1-\frac{i^2}{c^2}}\right); \qquad \text{7C1}$$

The intensity of the *neutron process* at the inflexion point:

$$\frac{dmc_s^2}{dt_n^s}\sqrt{1-\frac{(c_s-i_s)^2}{c_s^2}}\left(\sqrt{1-\frac{i_s^2}{c_s^2}}-1\right)>\frac{dmc^2}{dt_n}\sqrt{1-\frac{(c-i)^2}{c^2}}\left(\sqrt{1-\frac{i^2}{c^2}}-1\right); \qquad \text{7C2}$$

The intensity increase of the proton and neutron processes in 7C1 and 7C2 is also valid for any intermediate *v* speed values of the elementary processes of the impacted rod with loaded *Quantum Membrane*.

S.
7.1

7.1
Values of *Quantum Communication*

The value of quantum communication depends on the load of the *Quantum Membrane*. Empty space as such does not exist. *Space* itself is the *Quantum System of Reference*.

Energy quantum is consequence of quantum entropy generation, result of each of the cycles of the infinite chain of mass/energy and energy/mass transformations. Speed value *c*, the speed of quantum communication is changing, result of the load of the *Quantum Membrane*.

The *Quantum System of Reference* is subject to impacts.
Impacts load the *Quantum System of Reference* increase the intensity of mass/energy and energy/mass transformations, establishing with that the *Quantum Membrane*.
Impacts increase *c*, the speed established by energy and mass transformations. As *Quantum System of Reference* – the system of the energy quantum – does not accumulate energy, *Quantum Membranes* with different load communicate the information with different speed values.

These speed differences might be measureable, or might not be, as might be of infinite low values, but still considered as differences.

Ref.
S.8
S.9

The speed of quantum communication means the quantum space (*ds*) impact for unit period of time (*dt*). The actual load of the *Quantum Membrane*, all elementary processes of the micro and macro worlds, all our everyday activities influence and impact this speed value.

The *blue shift* of the electron process is impacting the *Quantum Membrane*. Proton and neutron processes communicate in line with the load and by the speed of the quantum communication of the *Quantum Membrane*.

Elementary processes at certain distance impact the quantum space; create local *Quantum Membrane* with increased quantum speed. The load of the *Quantum Membrane* depends on the intensity of the elementary processes. The distance and the increased quantum communication are measurable; they correspond to magnetic features of processes and elementary structures. All elementary processes have certain magnetic impact.

Elementary processes themselves might be subjects to impacts as well. The modification of the intensity of the local *Quantum Membrane* of the element in this case is proportional to the impact.

7.2
Magnetic impacts and features

With reference to 7C1 and 7C2, as result of external electron process *blue shift* impact, there is an increased intensity shift between the neutron and the proton processes of the elementary structure of the rod. The reason is that while ε_x the function of the intensity coefficient of the electron process remains the same;

$$\varepsilon_x = \frac{\varepsilon_p}{\varepsilon_n} \sqrt{1 - \frac{(c-i)^2}{c^2}}$$

the intensity of the *blue shift* impact, the acting energy intensity of the electron process and the *quantum communication* have been increased.

$e_{is} > e_i$ and $c_s > c$

7D1

Electromagnets or natural magnets are limited in their size, but most importantly, they have end surfaces. This is the place for the appearance of *magnetic features.*

The surfaces at the two ends of the rod are of *extraordinary* nature:
They do connect the *Quantum System of Reference* of the rod of increased quantum communication and increased *blue shift* impact with the surrounding environment of normal intensity conditions.

The appearance of magnetic features depends on the shift of the *intensities* of the integrated proton and neutron processes within the rod of the magnet.
The internal proton-neutron process balance of an element (or mineral) sensitive to magnetic functions is close to equilibrium status, still with neutron process intensity dominance (as having solid structures.)
> (External electron process *blue shift* impact flow from winded coil would generate similar reaction in all kinds of elementary rod structures just the appearances of the magnetic characteristics would be less notable.)

There are two ways of the communication of magnets through their two end-surfaces with the environment:
(1) at one end: proton process impact of increased intensity – as result of the increased speed of quantum communication; and
(2) at the other end: proton process demand also of increased intensity (as "neutron process impact").

The increased intensity shift between proton and neutron processes means increased absolute difference between proton and neutron processes in all sections of the internal elementary structure of the rod in its loaded status.
With reference to 7C1 and 7C2, the increased neutron-proton intensity shift can be demonstrated as:

$$A \cdot P > P; \ A \cdot N > N;$$
$$A \cdot N - A \cdot P = A \cdot (N-P) \quad \text{and} \quad A \cdot (N-P) > N - P$$

7D2

Proton and neutron processes go through the same time frames from t_o - the inflexion point to t_{is} - the electron process stage.

The absolute balance of the proton-neutron process is still valid:

7D3
$$\frac{dmc_s^2}{dt_p^s \varepsilon_p^s}\left(1 - \sqrt{1 - \frac{v^2}{c_s^2}}\right) = \frac{dmc_s^2}{dt_n^s \varepsilon_n^s} \xi \sqrt{1 - \frac{(c_s - i_s)^2}{c_s^2}}\left(\sqrt{1 - \frac{v^2}{c_s^2}} - 1\right);$$

The absolute balance in 7D3 is realised with increased intensity difference. The proton intensity cover demand of the neutron process is increased:

7D4
$$\frac{dmc_s^2}{dt_p^s}\left(1 - \sqrt{1 - \frac{v^2}{c_s^2}}\right) \neq \frac{dmc_s^2}{dt_n^s} \sqrt{1 - \frac{(c_s - i_s)^2}{c_s^2}}\left(\sqrt{1 - \frac{v^2}{c_s^2}} - 1\right);$$

Elementary process only happens if the neutron process, driven by electron process *blue shift* impact is covered by proton process belonging to the electron process drive. Whatever is the intensity increase of the internal *Quantum Membrane* of the rod, winded by wire with electron *blue shift* flow – if the proton cover is not provided, neutron process will not happen.

- At the *start surface* of the rod with electron process *blue shift* flow within the wire, the proton process cover is *limited*, since the neutron process is with more intensity. There is no *space* and *time* for the proton process to take from: the start surface in the direction outside the rod is without sufficient proton cover; in the other direction (within the rod) proton processes with increased intensity are in limited numbers, as the electron process *blue shift* impact only starts from this section.
  This is the reason the increased absolute proton process intensity cover need of the neutron processes cannot be provided. The difference in proton and neutron process intensities cannot be covered.
- The situation for all following surfaces and zones up to the end surface is consolidated: electron process *blue shift* of increased intensity within the rod can provide intensified proton process cover to the intensified neutron processes from the quantum space of the rod.
- The *end surface* is different again. Proton, electron and neutron processes are parts of the same cycle. Their absolute numbers are the same, just they are acting with intensity differences. Proton cover with less intensity has been left behind in time, covering the neutron process from the count of the previous sections of the rod. At the time moment the last neutron process of the rod is covered, there are still proton process intensities available in the rod. Therefore the last surface is with proton process *surplus*. (The missing proton process intensity at the start surface is granted de facto at the end surface.)

The increased difference in the intensities of the proton and neutron processes is valid in any space coordinates (sections) of the magnet in the direction of the *blue shift* impact from the coil.

The difference in proton and neutron process intensities has its impact on the measurement of the magnetic effect.

- big difference makes the effect non-measurable.
- no difference results in no effect.

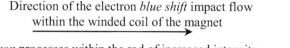

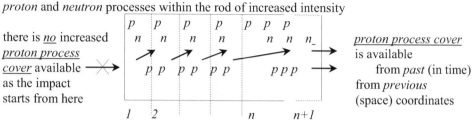

Fig. 7.2

Fig.7.2

The cover in *space and time* establishes a certain and permanent balance between neutron and proton process intensities starting from the start surface of the magnet with increased *demand in proton process intensity* towards the end surface with *proton process surplus and dominance.*

The proton process demand at the start surface and the proton process surplus at the end surface remain acting the same way in all cycles of the electron process *blue shift* impact flow of the magnet.

The established proton-neutron relations in *space* and *time* in the structure of the magnet results in permanent integrated intensity values alongside the rod from the start to the end surfaces.

From the start surface of the external *blue shift* impact onward, the intensity of the neutron process has been increased but the proton cover is maintained, as the sections of the rod starting from the initial electron flow impact through the following ones provide the increased proton process cover.

The integrated proton process intensities of the actual time frame and quantum space of the magnet, acting together with the ones from *"the past"* of the *"previously"* impacted *quantum space* keep balance with the neutron processes within the internal structure of the magnet, exempt the starting and the end surfaces.

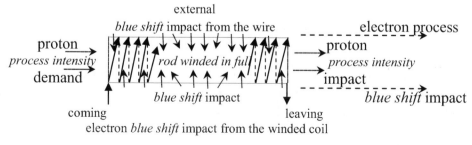

Fig.7.3

Fig. 7.3

> ➢ The integrated proton process intensity at the *end* surface of the external electron flow means increased proton process impact to the environment as there is no need for the proton cover at the end surface.
> ➢ At the *start* surface the relation is its opposite, the increased neutron process intensity is in proton process cover need, but missing it as the impact starts in time and space from here.

Surfaces at the start and at the end of the external impact of the rod also differ in their electron process intensities. There is not just extra not used proton process available at the end surface of the magnet, but electron process *blue shift* surplus as well, as electron process follows the proton process.

The situations at the start and at the end surfaces are unique:
-   The proton process impact surplus is leaving the end surface and landing as cover at the neutron process of the other end, the start surface, while
-   the electron process, the end product of the proton process is remaining within the magnetic rod but acting as *blue shift* impact all around the quantum space (in classical magnet arrangements, towards the start surface of the magnet) finally turning into neutron process.
-   The increased neutron process intensity at the start surface can have proton process cover and electron process *blue shift* drive only from the end surface of the rod with established electron process surplus and proton process cover available.

Magnets are the real proof of the *Strong Interrelation* demonstrating unity!

The *Strong Interrelation* at the end surfaces of magnets is acting do we realise it or not. In the case of close-in-space relations the effect is visible and very intensive.
If the intensity relations of the elementary processes within the magnet and in other objects in the surrounding environment are similar, elementary structures can be *attracted* to provide the necessary proton cover to the neutron process demand or can be *repelled*, result of the conflict between proton process intensities acting in parallel.

The load of the *Quantum Membrane* in magnets is increased, quantum communication is of higher speed, the time system corresponds to the speed value, the intensity of the electron *blue shift* impact is of high value.

While there is also electron process *blue shift* surplus at the end surface, magnets demonstrate in fact the *Strong Interrelation*.
Proton and neutron processes have become increased the same way, the function of the elementary intensity relations (ε_x) remained unchanged.

The cylindrical surface of the magnet is part of the balance of the internal sections. The integrated and simultaneous proton process ensures the cover for the neutron processes of the external winded surface as well. Therefore magnets do not show magnetic features (attracted or repelled) alongside this surface.

Natural magnets are representing more intensive quantum communication with the environment: elementary processes (proton, neutron and electron processes as well) are of higher intensity. In the case of infinite number of elementary processes, internal quantum communication and external relations are controlled by *Nature.*

Natural magnets are elementary structures with more intensive proton-neutron process balance and extra electron process *blue shift* drive (cause and effect are connected) with increased speed of quantum communication, communicating with the surrounding environment of normal intensity. They are of increased and directed intensity and power.
Magnets connect different time systems.

If *natural magnets* are taken (or cut) out from their natural acting balance in the environment, they react in similar ways: Sphere symmetrical release of energy in this case would not work. Internals cannot be emptied. The only way it can be controlled: circulation of the power, which in fact is at speed c_s time dt_p^s, dt_n^s and dt_{is} simultaneously acting through the end surfaces in unified quantum space.

As earlier discussed, all elementary structures can be intensified by electron process *blue shift* impact. Magnetic features however can only be expected from those minerals or elementary compositions, where the proton-neutron process balance is close to equilibrium (as indicator of the sensitivity of the elementary relations), still with neutron process dominance.
These are the basics of the operation of an *electro-magnet*:
- Magnetic function is about elementary processes with increased speed value (c_s) and intensity of the internal quantum communication!
- The intensity of the electron process of the magnet (ε_x) is without change.
- The two end-surfaces connecting the time systems and the intensities of the environment and the electro-magnet – have balance deviations:
  o the end surface is with proton process intensity and electron process *blue shift* surplus;
  o the start surface is in proton process intensity and electron process *blue shift* demand.
- Start and end surfaces of electro-magnets communicate and ensure the continuity of the elementary process.

Ref
S.8. In the case of natural magnets the *space-time* (the intensity of the inflexion point, with reference to Section 8) of the mineral – as result of external *blue shift* impact – is not homogenous. The intensity of the inflexion point is changing in a certain direction. The reason is that during the creation of the mineral, independently is it a single element or a mix of elements, external *blue shift* impact (heat or pressure) has influenced the process in this direction and had impact on the internal quantum structure of the mix.

The proton-neutron process balance within the internal structure of the mix has not been influenced, as the continuity of the mineral provides the *blue shift* and proton cover as the mineral needs it, but at the two surfaces the continuity of the cover is disrupted: there is neutron process intensity surplus at one end, while the other end is with proton process intensity and electron process *blue shift* surplus.

S.
7.3

7.3
Electricity flow within a single wire

Electricity flow is electron process *blue shift* impact transfer alongside within the wire. As result of the impact, the normal electron process *blue shift* impact within the elementary structure of the wire has been impacted from external source. This is not an electron (as particle) flow rather electron process *blue shifts* impact flow, provided by (from) external source.

The incoming external *blue shift* impact is resulting in intensified internal electron process and increased (intensified) *blue shift* impact within the wire (as it was within the winded rod as well) and taken further away by the *Quantum Membrane* of the wire (the flow).

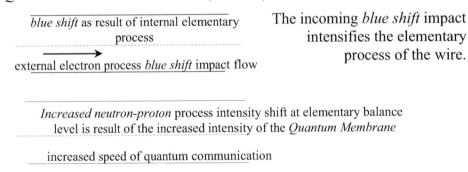

blue shift as result of internal elementary process

external electron process *blue shift* impact flow

The incoming *blue shift* impact intensifies the elementary process of the wire.

Increased neutron-proton process intensity shift at elementary balance level is result of the increased intensity of the *Quantum Membrane*

increased speed of quantum communication

Fig.
7.4

Fig. 7.4

The *blue shift* impact, arriving with the "electricity flow" is acting not just within the geometry of the wire. The intensified impact and conflict are affecting the *Quantum System of Reference* surrounding the wire as well. In conventional physics this impact (force) is called as magnetic *B-field*. This *loss* is one of the reasons why that the transported *blue shift* impact is always less than the incoming.

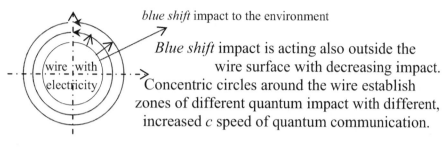

blue shift impact to the environment

Blue shift impact is acting also outside the wire surface with decreasing impact. Concentric circles around the wire establish zones of different quantum impact with different, increased *c* speed of quantum communication.

Fig. 7.5

Fig. 7.5

Concentric circles around the wire represent zones with equal speed of quantum communication. Different *Quantum Membrane* impact results in different quantum communication values. As the load of the *Quantum Membrane* around the wire is getting less and less with the growth of the distance from the surface of the wire, values of quantum communication (*c*) are with decreasing gradient, approaching the integrated value of the "speed of light" of the environment around the wire. The quantum processes in concentric circles around the wire transfer the *blue shift* impact with different intensities and quantum communication speed.

The incoming external *blue shift* impact flow in Fig. 7.4 does not drive the neutron process of the wire, since there is no proton process cover available to the incoming external electron process *blue shift* impact. The incoming *blue shift* impact generates *blue shift* surplus and conflict within the elementary structure of the wire. The result of the surplus and the conflict is *blue shift* impact flow (electricity) along the wire. This is the conflict and impact having effect to the environment surrounding the wire as well.

It is taken that external electricity load (quantum impact) to the wire results in n_s quantum impact at the internal surface of the wire. This impact should result in n_r quantum impacts around the wire. As the continuity of the *Quantum System of Reference* and the *blue shift* impact dictates, while the radius of the impact grows: $n_r > n_s$

The original absolute elementary *blue shift* impact within the wire without external load, equal to

$$E_x = \frac{dmc^2}{dt_i \varepsilon_i \varepsilon_x}\left(1 - \sqrt{1 - \frac{(c-i)^2}{c^2}}\right) = \frac{dn}{dt_i \varepsilon_i \varepsilon_x} q \qquad \text{7E1}$$

will be increased by the external load.
By the growing number of *quantum* space (by the growth of the radius) outside the wire the intensity impact of this absolute load will be decreasing.

7E2

$$e_r = \frac{dmc^2}{dt_i \varepsilon_x}\left(1 - \sqrt{1 - \frac{(c-i)^2}{c^2}}\right) = \frac{dn}{dt_i \varepsilon_x} q$$

Quantum is the entropy product of each of the infinite cycles of neutron-proton-neutron… transformation. The *Quantum Membrane* within the wire is overloaded. Overload means increased intensity and frequency.

The question is what is the description of the absolute value of the overload at the internal surface of the wire?

The increased load means *blue shift* conflict: more frequent quantum communication within the *Quantum Membrane*.

Should this indeed mean the increase of *c*, the speed of quantum communication?

The number of elementary processes within the wire stays as earlier, before the external impact. The incoming external *blue shift* impact is however extra load to the same quantum processes.

Quantum System of Reference is infinite and the *Quantum Membrane* does not accumulate energy intensity!

If there is any membrane function, the result of the increased load should mean increased communication; otherwise the *Quantum Membrane* would accumulate the impact.

Ref.
7A5
7A6

Therefore more intensive quantum communication means more quantum impacts with higher i and c speed values, with reference to 7A5 and 7A6 of the corresponding time system of specific $i = \lim c$ speed relation and with negligible time effect!

Can ε_x the function of the intensity coefficient of the electron process of the element be of different value?

It cannot be, as the elementary structure remains the same.

External *blue shift* load without proton process cover does not change the elementary structure.

With reference to 7E2 the intensity at the internal surface of the wire is:

7E3

$$e_s = \frac{dmc_s^2}{dt_{is} \varepsilon_x}\left(1 - \sqrt{1 - \frac{(c_s - i_s)^2}{c_s^2}}\right) = \frac{dn_s}{dt_{is} \varepsilon_x} q$$

dn_s means more quantum communication, while the total number of quantum within the space available within the wire remains n unchanged.

With reference to 7E2 and 7E3 the increase can be related to the speed of quantum communication and to the number of quantum impacts.

It is taken: $\left(1 - \sqrt{1 - \frac{(c_s - i_s)^2}{c_s^2}}\right) = \left(1 - \sqrt{1 - \frac{(c_r - i_r)^2}{c_r^2}}\right)$

$R(c^2) = dn$.   (*R* represents a specific function.)

This way: $R(c_s^2) - R(c_r^2) = dn_s - dn_r$; and 7E4

$$c_s^2 - c_r^2 = \frac{\Delta n}{R} = (c_s - c_r)(c_s + c_r) \quad \text{or} \quad \Delta n = R(c_s - c_r)(c_s + c_r)$$ 7E5

For the open *Quantum System of Reference* without load: $c_r = c$ 7E6

As result of the propagation of the generated within the wire increased quantum impact – with the growth of the radius of the impact the quantum number increases. As consequence, the intensity of the impact is obviously decreasing. If the increase still can be covered by the increased value of quantum communication generated by the external impact (electricity load) of the wire, the impact is acting. If not, the propagation of the impact stops.

The intensity of the impact is decreasing: $e_s > e_r$

The number of quantum communicating outside the surface of the wire is increasing by the growth of the radius. The intensity of the impact is changing and getting less and less with the growth of the distance from the surface of the wire.

7.4
Quantum impact of wire with electricity

S.
7.4

Blue shift impact flow (electricity) within a wire increases the intensity of the elementary process of the wire – loads the *Quantum Membrane.*
Loaded *Quantum Membrane* means, while there is no change in quantum numbers, impact/information/energy to be communicated, it is more than without load: The quantum communication is of increased speed.

A solenoid with wires with electricity flow intensifies the *Quantum Membrane,* increases the intensity of the elementary process of the rod of the solenoid or increases the *Quantum Membrane* of the *Quantum System of Reference* within the solenoid (without any internal rod).

The *blue shift* flow is impact to the *Quantum Membrane* within the wire as well. The intensification of the *blue shift* impact of the internal electron process of the wire intensifies the neutron collapse. The proton process as cover has been intensified through the intensification of the inflexion point. The quantum communication becomes of increased speed.

The *Strong Interrelation* and the continuity of the *Quantum System of Reference* rule the reaction of magnetic effects.

What is the direction of the magnetic effect if *Earth's* magnetic field is directed from the *North*-pole to the *South*-pole?

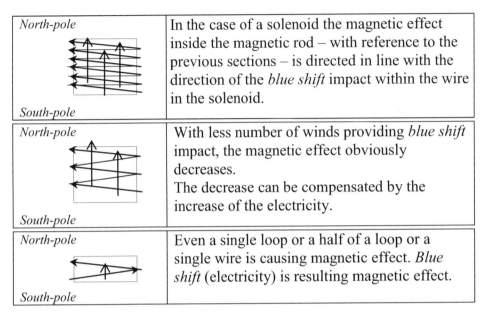

Fig.
7.6   Fig. 7.6

In the case of solenoid is without rod, just with the *Quantum System of Reference* within the internal space, *quantum communication* is directed the same way towards the direction of the winds of the solenoid.

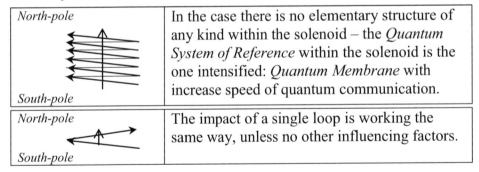

Fig.
7.7   Fig.7.7

The direction of magnetic lines within the solenoid is based on the continuity of the *Quantum System of Reference*. Once the load starts, the increased quantum communication creates certain proton process need with external quantum continuity demand, establishes *South*-pole.

Quantum communication and with that the power of the internal magnetic lines of the *Quantum Membrane* is directed towards the other end, the *North*-pole with *blue shift* impact surplus and proton process reserve.

In the case we open a single loop, increasing the radius to infinity – the wire will have its own magnetic lines. Quantum communication will follow the increased status of the *Quantum Membrane* around the wire. It will be created around the wire as concentric circle/s. The continuity of the *Quantum System of Reference* will ensure the circulation in concentric lines with changing speed as demonstrated in the previous section.

The direction of the magnetic lines in this case however is result of other factors as well.

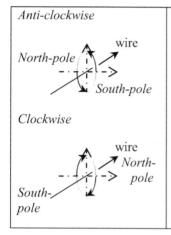	In the case of just a single wire with *blue shift* impact flow, the *blue shift* effect with increased quantum communication within the *Quantum Membrane* can be directed relative to the direction of the *blue shift* impact flow within the wire in two ways:

clockwise or anti-clockwise.

The direction of the quantum impact within the wire is based on the less intensive proton process. The intensified neutron process goes ahead with the *blue shift* flow.
The less intensive proton process follows and dominates the *Quantum Membrane.*

Fig. 7.8

Fig.
7.8

The influencing factors here are: the magnetic field of *Earth* and the *blue shift* impact of the sphere symmetrical expanding acceleration of the *Earth* (*gravitation*).

Magnetic lines above *Earth* surface are directed from the *North*-pole to the *South*-pole. The direction of the internal elementary *blue shift* flow of the *Earth*, deep inside below the surface is oriented from the *South*-pole towards the *North*-pole. The geographic *North*-pole provides *blue shift* and proton cover above the *Earth* surface towards the neutron process demand of the *South*-pole. The loop is closed and the response is given to the elementary system within the core of the *Earth*, the initiator of the processes.

At the same time *Earth* is in sphere symmetrical expanding acceleration at constant $i = \lim a\Delta t = c$ speed of our space-time system. As expanding acceleration, the surface of the *Earth* is in permanent impact with the *Quantum System of Reference*, resulting permanent *blue shift* impact (*gravitation*) and generating the *Quantum Membrane* on the *Earth* surface.

The expanding acceleration and the *blue shift* impact of the *Earth* are supplemented by the fact that *Earth* is rotating, responding to the *blue shift* impact of the *Sun*.

The two *blue shift* impacts (of the *Sun* and the *Earth*) result in the rotation of the *Earth* as working out the conflict. [*Sun's blue shift* is originating from the elementary *blue shift* conflict of the *Helium* and the *Hydrogen*.

The intensity of the neutron process of the *Hydrogen* is of infinite low value and *Hydrogen* generates permanent *blue shift* impact. *Helium* is with *blue shift*, surplus, as it is with proton process intensity dominance.]

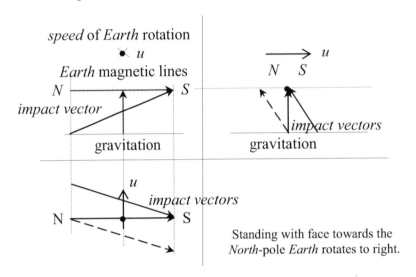

Fig.
7.9

Fig. 7.9

There is a certain impact vector initiating from these natural factors at each geographic coordinates of the *Earth*. This impact vector is directing the quantum communication either *clockwise* or *anti-clockwise* relative to the direction of the electricity flow of the wire.

[In the case of the *Ampere*-loop the *magneto-static dipole* the direction corresponds to right hand rule.]

In the case of gaseous or liquid "cables" for transferring *blue shift* impact, the direction would be the opposite, as in gaseous elementary state the proton process is the more intensive one. In gaseous state the direction of the magnetic effect is the opposite and the neutron process is dominating the magnetic field.

7.5
Electro-magnetic motors

If a wire is taken between the two poles of a magnet – but at rest and without electricity flow within the wire – there is no conflict between the *blue shift* drive of the magnet and the elementary process of the wire.

The *Quantum Membrane* of magnet with increased speed of quantum communication provides *blue shift* drive and proton process cover to the neutron process of the *South*-pole from the *North*-pole of the magnet.
Elementary relations within the wire correspond to their normal standards.

The *blue shift* drive and the *Strong Interrelation* of the wire and the magnet operate within different time systems and with different speed values of quantum communication.

Different time systems and quantum communications with different speed values also mean different quantum space relations.

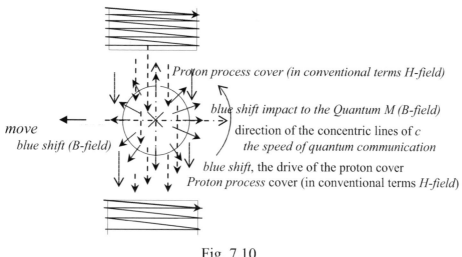

Fig. 7.10

Fig. 7.10

If the wire moves in any direction different than that of the direction of the *Strong Interrelation* of the magnet, the *blue shift* impacts of the elementary processes of the wire become conflicting and interrupting the *Strong Interrelation* of the two poles of the magnet.

While the interruption happens, which is rehabilitating immediately – as there are no elementary process without acting *Strong Interrelation* – this is an impact to the *Quantum Membrane* of the element of the wire from the "external source" of the magnet as well, generating additional *blue shift* impact. The generating additional *blue shift* impact "flows" away within the wire as it is without proton process cover.

The increasing intensity of the *Quantum Membrane* within the wire results in increased speed of quantum communication around the surface.

The direction of the move of the wire shall correspond to the direction of the *Strong Interrelation.* The other direction is the conflict of quantum communications of the side pushing the wire away.

The establishing *clockwise* or *anti-clockwise* direction of the quantum communication of the wire corresponds to acting natural conditions and this way identifies the direction of the electricity flow.

In the case the wire between the end surfaces of an electro- (or natural) magnet is connected to an external electricity supply, the propagating *blue shift* flow loads the *Quantum Membrane* within the wire.

The increased *Quantum Membrane* results in increased and changing quantum communication around the wire. The direction of the concentric cycles with equal speed of quantum communication around the wire will be corresponding to the acting natural effect of *Earth gravitation* and *Earth magnetic lines*, with reference to Figure 7.10.

With reference to Fig.7.10, the side (region) of the magnet-wire relation with conflicting quantum communication (where the speed directions are opposite) is pushing the wire into the opposite direction with speed of quantum communications of similar or parallel directions.

> Electricity flow within the wire results in similar symptoms as magnetic rod has under solenoid impact, but internal magnetic features within the wire cannot be noted, as they move ahead together with the *blue shift* impact, generating them. Magnetic impact can be measured at the surface.

7.6
Magnetic minerals

The format of the overall elementary balance is:

$$\frac{dmc^2}{dt_p \varepsilon_p}\left(1-\sqrt{1-\frac{v^2}{c^2}}\right) = \frac{dmc^2}{dt_n \varepsilon_n}\xi\sqrt{1-\frac{(c-i)^2}{c^2}}\left(1-\sqrt{1-\frac{v^2}{c^2}}\right)$$

Elementary balance is based on quantum communication between elements with electron process *blue shift* surplus and *blue shift* deficit.

The specifics of this communication is that while *blue shift* surplus means electron processes more in numbers and less in intensities, *blue shift* deficit is with electron process demand but of higher intensity.

This way the integrated *Quantum Membrane* becomes being loaded on electron processes in numbers and intensities, between the values of the components.

$$\frac{dmc^2}{dt_{iA} \varepsilon_{xA}}\left(1-\sqrt{1-\frac{(c-i)^2}{c^2}}\right) < \frac{dmc_n^2}{dt_{int} \varepsilon_{x\,int}}\left(1-\sqrt{1-\frac{(c_i-i_i)^2}{c_i^2}}\right) < \frac{dmc^2}{dt_{iB} \varepsilon_{xB}}\left(1-\sqrt{1-\frac{(c-i)^2}{c^2}}\right)$$

$\quad$ load of element A $\qquad\qquad$ load integrated $\qquad\qquad$ load of element B

The load of the *Quantum Membrane* determines c, the speed of quantum communication and dt_i the time system of the integrated electron process.

If elements or compounds are subjects to external *blue shift* impact without elementary communication (without proton-neutron process relations) the *Quantum Membrane* becomes of increased load. The explanation is that the direct *blue shift* impact does not increase the quantum space, instead results in *blue shift* conflict.

Elementary components bring their *Quantum System of Reference* as *Quantum Membrane* (quantum space) into the relation.

Blue shift conflict increases the load of the *Quantum Membrane*:

$$\frac{dmc^2}{dt_{int}\varepsilon_{x\,int}}\left(1-\sqrt{1-\frac{(c-i)^2}{c^2}}\right)+\frac{dmc_n^2}{dt_{in}\varepsilon_{xn}}\left(1-\sqrt{1-\frac{(c_n-i_n)^2}{c_n^2}}\right)>\frac{dmc^2}{dt_{int}\varepsilon_{x\,int}}\left(1-\sqrt{1-\frac{(c-i)^2}{c^2}}\right) \qquad 7G3$$

integrated load   additional *blue shift*

The original *Quantum Membrane* is of increased intensity and load.
Blue shift drive of increased intensity results in neutron process of increased intensity with natural need for increased proton process cover.

Magnets have *Quantum Membrane* of increased load. This increased load can be homogenous or directed. Meaning: the intensity is the same in any direction or the intensity vector has a certain direction.

While fixing the direction of the *blue shift* impact in electromagnets is easy, the explanation in natural magnets is more complex.

Gravitation as electron process *blue shift* impact is of quasi permanent value and direction.
Elements and elementary compounds vary.
All elements and compounds communicate with *gravitation*.
The actual elementary behaviour however depends on the quality of the electron process *blue shift* impact of the elementary composition.
    Elements with electron process *blue shift* surplus and this way with certain load of the original *Quantum Membrane* cannot be impacted.
    To influence elements with deep electron process *blue shift* deficit and therefore with increased electron process intensity is also difficult.

This separates elements and compounds being diamagnetic, paramagnetic, ferromagnetic, ferrimagnetic, "antimagnetic" etc.

Natural magnetism means *gravitation* is impacting the *Weak* and the *Strong Interrelations* of elementary processes.

Hydrogen, Oxygen, and *water, Nitrogen, Carbon, Silicon, Calcium, Sulphur* and *Helium* are *diamagnetic.* Compounds of these elements with others close to equilibrium state, like *Magnesium, Aluminium, Sodium, Potassium, Titan* might be *paramagnetic.*
Minerals of *Iron, Cobalt* and *Nickel* in certain proportions with other elements are *ferromagnetic.* What makes a compound to ferromagnetic?

Fe	0.88347	These elements have different electron process
Co	0.85734	intensity coefficients. In certain mineral compositions
Ni	0.92498	they all are with solid ferromagnetic features.

Composition of elements with integrated elementary balance close to equilibrium state, but below it are very sensitive to external impacts, while others, while experiencing the impact have no effect.

A ferromagnetic elementary composition is with sensitivity (susceptibility) good enough for being effected by *gravitation*. The impact of *gravitation* is directing the proton-neutron relation. The neutron collapse and its proton demand have its certain direction.

In the case of composition of elements with integrated proton and neutron process intensities at quasi equilibrium state *gravitation* is an additional, but decisive *blue shift* impact to the *Quantum Membrane* of the composition.
As the intensities of the driven neutron processes and the covering proton processes are at quasi balance, (which is exclusively rear for intensities) the direction of *gravitation* has its significant impact to the process.

The *Quantum Membrane* of ferromagnetic minerals is composed from the *blue shift* impacts of *Fe* and element *X* and the *blue shift* impact of *gravitation*.

7H1

$$\frac{dmc^2}{dt_p^{Fe}}\left(1-\sqrt{1-\frac{i^2}{c^2}}\right)+\frac{dmc^2}{dt_p^{X}}\left(1-\sqrt{1-\frac{i^2}{c^2}}\right)\cong$$

$$\cong\frac{dmc^2}{dt_n^{Fe}}\sqrt{1-\frac{(c-i)^2}{c^2}}\left(1-\sqrt{1-\frac{i^2}{c^2}}\right)+\frac{dmc^2}{dt_n^{X}}\sqrt{1-\frac{(c-i)^2}{c^2}}\left(1-\sqrt{1-\frac{i^2}{c^2}}\right)$$

The close to equilibrium status proton-neutron intensity balance means, the proportions of the proton/neutron communications of the two elements are established close to each other.
The minor effect of the *blue shift* of *gravitation* therefore has key impact at both, the elementary processes and in addition at their *direction*.

7H2

$$(Fe)\frac{dmc^2}{dt_i\varepsilon_{Fe}}\left(1-\sqrt{1-\frac{(c-i)^2}{c^2}}\right)+(X)\frac{dmc^2}{dt_i\varepsilon_{X}}\left(1-\sqrt{1-\frac{(c-i)^2}{c^2}}\right)+$$

$$+(G)\frac{dmc^2}{dt_g}\left(1-\sqrt{1-\frac{(c-i)^2}{c^2}}\right)=\frac{dn}{dt}q$$

The driving proportions of the electron processes of (*Fe*) and (*X*) will be function of value of (*G*), the impact from *gravitation*.

Since this impact is with certain direction, the *Strong Interrelations* for *both* elements will be established in line with the direction of *gravitation*.
This means: the neutron process drive and the proton process cover have their certain significant direction, the direction of the sphere symmetrical expanding acceleration of *Earth*.

The driving force of elementary communications is looking for balance.
In this certain *Fe-X* case the balance is not just about the absolute but also about intensity values.

The internal structure of ferromagnetic minerals is homogenous: the balance is established at any level of the elementary structure including the surface.

> The *Strong* and *Weak Interrelations* of the internal elementary structure of minerals have the balance with the surrounding elements of the natural environment.

The mining out ferromagnetic minerals from their natural environment does not change their magnetic features: Internal elementary processes continue to happen the same way (without the definite direction of *gravitation*) as the internal communication follows the demand of the neutron collapse (in line with the earlier established routine).

This also means that the boundaries of ferromagnetic minerals with proton process cover demand and surplus (as established by the impact of *gravitation*) remain the same.

In order to change the built in *Weak* and *Strong Interrelation* ferromagnetic minerals shall be heated up. Heating is additional *blue shit* impact which overrules the direction established by *gravitation*. Heating is an impact with direction as well, but the effect is so strong that the resulting *blue shift* conflict becomes "chaotic".

8

S.8   *Space-time and quantum gradient*

There are three key processes in the establishment of the *Quantum System of Reference*:
-   *quantum entropy* generation;
-   *inflexion point*;
-   generation and accumulation of the *energy quantum*.

8A1   *Quantum entropy* is generating at the *blue shift* level of the electron process, as $qe = \dfrac{dmc^2}{dt_i}\sqrt{1-\dfrac{(c-i)^2}{c^2}}$

8A2 $e_{iqe} = e_i - (e_i - e_{i1}) - ... - (e_{in-2} - e_{in-1}) - \left(\dot{m}_{in-1}c^2 - \dot{m}_{in}c^2 \sqrt{1-\dfrac{(c-i)^2}{c^2}} \right) = \dot{m}_{in}c^2 \sqrt{1-\dfrac{(c-i)^2}{c^2}}$

Quantum entropy is the last component of the chain of electron processes. The last component with $\lim a = 0$ and $\lim \Delta t = \infty$ parameters.

8A3 $e_e = \dfrac{dmc^2}{dt_i \varepsilon_x}\left(1-\sqrt{1-\dfrac{(c-i)^2}{c^2}} \right) = \dfrac{dmc^2}{dt_i \varepsilon_x}\left(1-\sqrt{1-\dfrac{(a_0 \Delta t_\infty)^2}{c^2}} \right);$

And ε_x the intensity coefficient of the electron process, equal to $\varepsilon_x = \dfrac{\varepsilon_p}{\varepsilon_n}\sqrt{1-\dfrac{(c-i)^2}{c^2}}$

8A3 corresponds to the end stage of the neutron collapse, as $\lim a = 0$.
The neutron process is reaching the inflexion point with c, the speed of quantum communication.
ε_x is specific and distinguishing characteristic of each element.
[The less is the value of the intensity coefficient of the electron process, the higher is the intensity of the electron process.]

The *inflexion point* as endpoint and starting point is equally relating to the neutron collapse and to the proton expansion processes as well: The more intensive the neutron collapse – proton generation process is, the more intensive (shorter) is the inflexion point and with that the intensity of the *quantum* entropy generation.

8A4   Quantum entropy is the last energy/mass component of the *blue shift* chain of the electron process. $e_e = \dfrac{dmc^2}{dt_i \varepsilon_x}\left(1-\sqrt{1-\dfrac{(c-i)^2}{c^2}} \right)$

8A5   This way, the proton process generates absolute expansion energy/mass value, portion of $E_p = mc^2 \left(1-\sqrt{1-\dfrac{i^2}{c^2}} \right)$

The neutron process is getting back collapsing mass/energy absolute portion only of

$$E_n = mc^2 \sqrt{1 - \frac{(c-i)^2}{c^2}} \left(\sqrt{1 - \frac{i^2}{c^2}} - 1 \right) \qquad 8A6$$

Quantum energy is an energy intensity value of the expansion of the matter equivalent to the intensity difference of the proton and neutron processes, which leaves the *Strong Interrelation* of the elementary process:

$$q = e_p - e_n = \frac{dmc^2}{dt_o} \left(1 - \sqrt{1 - \frac{i^2}{c^2}} \right) - \frac{dmc^2}{dt_o} \sqrt{1 - \frac{(c-i)^2}{c^2}} \left(1 - \sqrt{1 - \frac{i^2}{c^2}} \right) \qquad 8A7$$

-   as proton process covers the neutron process, and the cover is permanent,
-   as the cover is an integrating impact of all processes running in parallel (meaning: the neutron, proton and electron processes represent three different cycles running in parallel in the infinite chain of processes)

the intensity difference of the cover and the demand at the inflexion point gives quantum energy value of:

$$q = \frac{dmc^2}{dt_o} \left(1 - \sqrt{1 - \frac{i^2}{c^2}} \right) \left(1 - \sqrt{1 - \frac{(c-i)^2}{c^2}} \right) \qquad 8A8$$

8A8 above is the quantum energy portion generating within the first elementary cycle, where the starting energy intensity was: $\dfrac{dmc^2}{dt_o}$ $\qquad$ 8A9

At the completion of cycle *n* the value of the energy quantum is

$$q_n = \frac{dmc^2}{dt_o} \left[\sqrt{1 - \frac{(c-i)^2}{c^2}} \right]^{n-1} \left(1 - \sqrt{1 - \frac{i^2}{c^2}} \right) \left(1 - \sqrt{1 - \frac{(c-i)^2}{c^2}} \right) \qquad 8A10$$

creating with the *Quantum System of Reference* = space!

All quantum energy values of all cycles are equal, as the threshold value of the *Quantum Membrane* against the electron *blue shift* is the same:

$$q_1 = q_2 = \ldots = q_{n-1} = q_n \qquad \text{and for all values equal: } \lim q_n = 0 \quad 8A11$$

➤   The time system of the generation of *quantum entropy* belongs to speed of $i = \lim a\Delta t = c$:

$$qe = \frac{dmc^2}{dt_i} \sqrt{1 - \frac{(c-i)^2}{c^2}} = \frac{dmc^2}{dt_o} \sqrt{1 - \frac{i^2}{c^2}} \sqrt{1 - \frac{(c-i)^2}{c^2}} \qquad 8B1$$

This intensity at the inflexion point corresponds to:

$$qe_o = \frac{qe}{\sqrt{1 - \frac{i^2}{c^2}}} ; \qquad \text{(an intensity of infinite time more)}$$

to $\qquad qe_o = \frac{dmc^2}{dt_o \sqrt{1 - (i^2/c^2)}} \sqrt{1 - \frac{i^2}{c^2}} \sqrt{1 - \frac{(c-i)^2}{c^2}} = \frac{dmc^2}{dt_o} \sqrt{1 - \frac{(c-i)^2}{c^2}} \qquad 8B2$

Ref
Book2
S.14

> *Energy quantum* = quantum is generating at speed level $i = \lim a\Delta t = c$ as above in 8A6, and leaves the elementary process by c, the speed of quantum communication.

14A8   The transformation during proton process expansion can be described as

8B3

$$d\left(m_o c^2 - m_v c^2\right) = \frac{dp_{tr}}{dt_{tr}} ds_{tr}; \quad \text{and} \quad d\left(m_o c^2 - m_v c^2\right) = \frac{d(m_{tr}\upsilon)}{dt_{tr}} ds_{tr}$$

where dp_{tr} – is the momentum of the proton mass transformation;

dt_{tr} and ds_{tr} – are the duration and the path of the proton mass

transformation;

m_{tr} – is the value of the transformed mass; and

υ – is the velocity of the mass transformation, the speed value of the expansion ("disappearance") of mass.

$\upsilon \neq v$ or $\upsilon \neq \Delta v$ as v or Δv are the speed or the speed difference between the stages of the sphere symmetrical expanding acceleration of the proton process. At its end stage $v = i$.

8B3 means, the mass difference of the missing proton stage transforms into quantum energy: Proton process expands the space and creates the *Quantum System of Reference*.

Can the value of υ in 8B3 be other than c?

8B4   Resolving the equation, 9B3

gives: $d\left(m_o c^2 - m_v c^2\right) = \left(\frac{dm_{tr}}{dt_{tr}}\upsilon + \frac{d\upsilon}{dt_{tr}}m_{tr}\right)\upsilon dt_{tr}$

We suppose that the speed of the mass transformation is constant in time:

$$\frac{d\upsilon}{dt_{tr}} = 0 \qquad \text{(The result will prove that this condition is taken correctly.)}$$

8B5   8B3 gives: the integral of which results in:

8B6

$$\frac{d\left(m_o c^2 - m_o c^2 \sqrt{1-(v^2/c^2)}\right)}{dm_{tr}} = \upsilon^2 \qquad m_{tr} = m_o \frac{c^2}{\upsilon^2}\left(1-\sqrt{1-(v^2/c^2)}\right)$$

8B7

8B6 proves; there is only one valid option: $\upsilon = c$

The value of υ, the speed of the "disappearance" of the mass into energy – the speed of quantum generation *cannot be either more or less than c*.

The generation of *space* means an expansion by the speed of quantum communication. The expansion of the proton surplus (response to quantum entropy) – creates *The Space*.

Space is the collection and accumulation of energy components, "the remaining proton process surplus", speed of c.

Quantum communication is speed of c.

Space is part of our elementary quantum world!

Space is not about emptiness!

Space is equal to the *Quantum System of Reference*!

Space is everywhere as time exists everywhere!

As there is no time without event and there is no event without time there is in parallel and similarly no *space* without event and there is no event without *space*!

Quantum space has its own intensity value and transfers quantum impact by c the speed of quantum communication. As definition it is better to say: quantum speed is equal to the speed of the creation of the quantum energy.

The intensity of the quantum space can be increased: this is the *Quantum Membrane*, loaded by *blue shift* impact of electron processes.

Energy quantum (quantum) cannot exist at status of rest.

Status of rest or any decrease of c the speed of quantum communication and generation would mean that energy/mass might disappear and the energy/mass balance of the nature might be destroyed, which is impossible.

Energy quantum is the natural response to the quantum entropy:

The absolute values of the mass-energy proton expansion and the energy-mass neutron collapse of elementary cycles are never exactly equal.

Energy/mass quantum makes the balance perfect.

Energy/mass quantum represents process and intensity.

The mass value of the cycle at the inflexion point, passing through belongs to both of the half cycles, the neutron process as end and the proton process as start as well. With reference to 8B2, the entropy product is equal to this value at the inflexion point and is the appearance of two opposite but balanced *quantum process gradients,* acting as:

-    the gradient of the (mass) *collapse*: $-\dfrac{dmc^2}{dt_o}\sqrt{1-\dfrac{(c-i)^2}{c^2}}$ $(=qe_-)$ 8B8

-    the gradient of the (energy) *expansion*: $+\dfrac{dmc^2}{dt_o}\sqrt{1-\dfrac{(c-i)^2}{c^2}}$ $(=qe_+)$ 8B9

Quantum entropy gradients are processes, the two sides of the inflexion point. Quantum entropy gradients connect energy quantum with the proton/neutron mass-energy balance, the subject of Section 9.

Ref
S.9

S.
8.1

8.1
Quantum entropy gradients

Quantum entropy generation is permanent and natural part of the infinite cycle of elementary processes. Space is expanding in time.

Time is relativistic and our space measurements vary.

The *Quantum System of Reference* (the accumulating quantum space) is under permanent impact of quantum entropy gradients at the inflexion point. The time system and the speed of quantum communication establish the unity of the two *quantum* entropy gradients.

Quantum gradients provide a kind of *blue shift – red shift* balance,
originated at the inflexion point, establishing the quasi *impulse* status of the
Quantum System of Reference:

8C1   *Quantum* gradients are the two sides of the
same process and compose the absolute
value of the *quantum entropy* at the
inflexion point:

$$qe_o = \frac{2\left|\dfrac{dmc^2}{dt_o \varepsilon_o}\sqrt{1-\dfrac{(c-i)^2}{c^2}}\right|}{2}$$

8C2
$$qe_o = \left|\frac{qe_- - qe_+}{2}\right| = \left|\frac{qe_+ - qe_-}{2}\right| = \frac{dmc^2}{dt_o}\sqrt{1-\frac{(c-i)^2}{c^2}}$$

The quantum gradient is the difference of the endpoints of the change divided by 2.

Quantum entropy gradient (or just quantum gradient) keeps the *Quantum
System of Reference* under permanent impulse-impact, character of *blue shift
– red shift* effect, giving by this to the *quantum* of the *Quantum System of
Reference* permanent intensity load impact.

Blue shift load and intensity increase of the *Quantum Membrane* increases
the intensity of the inflexion point and the intensity of the quantum entropy
gradients. Why and what way?
Quantum entropy is developing at $i = \lim a\Delta t = c$.
The intensity increase of the *Quantum Membrane* means, electron process
blue shift is of more intensity:

8C2 $$e_e = \frac{dmc^2}{dt_i \varepsilon_x}\left(1 - \sqrt{1 - \frac{(c-i)^2}{c^2}}\right)$$

The only way to increase e_e is
increasing c, which means increased i
with corresponding time system dt_i.

$\lim(c - i) = const = 0$ and with reference to 8C4 ε_x does not have impact.
The increased speed value of quantum communication increases the
intensity of the quantum entropy gradient at the inflexion point.

8C3 $$\frac{dmc^2}{dt_o}\sqrt{1-\frac{(c-i)^2}{c^2}} < \frac{dmc_x^2}{dt_o}\sqrt{1-\frac{(c_x-i_x)^2}{c_x^2}}$$

While the inflexion point is one and the same, the relation of ds_o to dt_o is
changing: $ds_{ox} > ds_o$ and $c_x > c$
There is only one way for increasing the intensity of the quantum entropy
gradient: increasing the speed of quantum communication.
The increased intensity of the inflexion point means more than just the
increased intensity of the neutron-proton transition. This intensity is
establishing the speed value of the quantum communication
-    as quantum gradient is impacting the quantum content of the *Quantum
     System of Reference* (loading the *Quantum Membrane*) and
-    establishing the *space-time* relation of elementary processes.
The intensity of the inflexion point (and with that the speed of quantum
communication) are the basics of the *space-time* of elementary processes.

Space-time includes:

$$i_{var} = \lim a\Delta t = c_{var}$$ the time system and the speed value of the electron process are variant as well.

The importance of the inflexion point is obvious. It not just proves that c, the speed of quantum communication is variant but also demonstrates its close relation to the inflexion point.

In other words:

(1) the speed of quantum communication and the intensity of the inflexion point are interdependent;

(2) while the meaning of dt_o the inflexion point is one and the same – the speed of quantum communication establishes the *space-time*.

The intensity of the electron process remains unchanged	$\varepsilon_e = \dfrac{\varepsilon_p}{\varepsilon_n}\sqrt{1 - \dfrac{(c-i)^2}{c^2}}$	as $\lim(c-i)=0$ remains the same and infinitely small in any circumstances.

8C4

[The constant intensity value of the electron process means: constant characteristic function for each element, changing in time with the progress of the elementary cycle.]

The speed of quantum communication and the acting quantum gradient of the inflexion point establish the dynamism and the balance of the *Quantum Membrane*.

External *blue shift* impact further loads the *Quantum Membrane*. The intensity of the quantum entropy gradient in this case is further increasing.

The increase of c to c_s justifies the increase. dt_o - the inflexion point is the same, time counts belonging to i_s to i are quasi equal.	$\dfrac{dmc_s^2}{dt_{is}\varepsilon_{is}\varepsilon_x}\left(1 - \sqrt{1 - \dfrac{(c_s - i_s)^2}{c_s^2}}\right) > \dfrac{dmc^2}{dt_i\varepsilon_i\varepsilon_x}\left(1 - \sqrt{1 - \dfrac{(c-i)^2}{c^2}}\right)$
	and $\quad dt_{is} = \dfrac{dt_o}{\sqrt{1 - \dfrac{i_s^2}{c_s^2}}} \approx \dfrac{dt_o}{\sqrt{1 - \dfrac{i^2}{c^2}}} = dt_i$

Space-time represents the dynamic balance of the speed of quantum communication and the internal and external *blue shift* impacts.

Neutron collapse is driven by electron process *blue shift*.

If the *blue shift* impact is of increased intensity, the quantum communication of the neutron and proton processes, reference to 8C2 and 8C3, and the inflexion point as well are also of increased intensity.

Ref
8C2
8C3

In the case the intensity of the inflexion point is infinite high, the *space-time* system is approaching *zero* and the speed of quantum communication is approaching *infinity*!	$\lim \varepsilon_o = \lim \dfrac{1}{dt_o} = \infty$	8C5
	$\lim dt_o = 0; \quad \lim c = \infty$	8C6

Space-time does not exist without *quantum* (and *quantum* entropy).

Space-time is the accumulating energy *quantum* of elementary processes.

Space-time is the *Quantum System of Reference,* the existence of the matter, characterising intensity, quantum communication, *time* and *space* relations. *Quantum Membrane* corresponds to its loaded status.

Quantum withstands *blue shift* impacts, transferring them without change, without accumulating any energy or mass load.

Quantum entropy gradient is the origin of the speed of the impulse transfer of the quantum communication of the *Quantum Membrane*, which might further be loaded by external electron *blue shift* impact.

The load of *energy quantum*, result of the quantum entropy gradient impact varies, as the speed of quantum communication varies as well, and as being of entropy generated value – is accumulating!

Blue shift surplus or deficit, the balance of proton and neutron processes determines the physical state of elements within certain *space-time*.

Heat, pressure, temperature and "energy" increase in conventional terms are direct effects of the intensity of the *blue shift* impact.

Increasing quantum generation of the same intensity expands the space.

Space-time without elementary processes and this way without *blue shift* impact and load is losing on quantum entropy gradient intensity, on *c*, the speed value of quantum communication.

S.
8.2

8.2
No *space-time* without *quantum*

With the load of the *Quantum Membrane* the speed of quantum communication is increasing.

(The speed of quantum communication is measured as $c \cong 300000$ km/sec in our natural environment on the surface of the *Earth*. This is the speed value of the quantum communication on the surface of the *Earth*. The surface of the *Earth* is in speed $i = \lim a\Delta t = c$, while the *Earth* itself is expanding with $g = a = 9.81...$ m/sec acceleration.)

Space-time with elementary processes acting within represents always certain *Quantum Membrane* load status. The measured by us characteristics (mainly temperature) of the *Quantum Membrane* this way depend on the internal *blue shift* conflict of elementary processes within the quantum *space*.

In the case of no elementary processes (within the *space-time*) the *Quantum System of Reference* is without any *blue shift* impact, neither as generating within, nor as transferring it through as from external impact.

8E1   No impact as such is equivalent to a single impact for infinite time:
$$\left| \frac{dmc^2}{dt_o} \sqrt{1 - \frac{(c-i)^2}{c^2}} \right|; \quad \text{and} \quad \lim dt_o = \infty$$

8E2   Absolute vacuum is therefore of quasi *zero* absolute temperature: $\lim T = 0$

External *blue shift* impact can load the *Quantum Membrane* of quasi *zero* intensity, *infinite time* system and $\lim T = 0$ temperature.

More external load to the *Quantum Membrane* within limited *space* $ds = const$ is increasing c	$c_x = \dfrac{ds_x}{dt_o} > \dfrac{ds_o}{dt_o} = c_o$	equivalent to $ds_x > ds_o$
More internal space without increasing quantum impact is decreasing c	$c_x = \dfrac{ds_x}{dt_o} < \dfrac{ds_o}{dt_o} = c_o$	equivalent to $ds_x < ds_o$
Constant space with constant quantum impact but with increasing time period of the inflexion is decreasing c	$c_x = \dfrac{ds_x}{dt_o} < \dfrac{ds_o}{dt_o} = c_o$	equivalent to $ds_x < ds_o$
Constant space with constant quantum impact but with decreasing time period of the inflexion is increasing c	$c_x = \dfrac{ds_x}{dt_o} > \dfrac{ds_o}{dt_o} = c_o$	equivalent to $ds_x > ds_o$

Table 8.1

Tab. 8.1

Table 8.1 contains important data as it shows that the relation of the parameters of *space* and *time* or in other words the relative duration of the *inflexion* of elementary processes determines the intensity of the *space-time* and the speed of quantum communication. This is the intensity of energy quantum transfer impacts, by the speed of quantum communication.

It is important to note that the intensity of a single elementary impact and the integrated intensity of a *Quantum Membrane*, result of increased number of *blue shift* impacts are identical in their physical terms.

S. 8.2.1

8.2.1 Specifics of the Hydrogen

The *space-time* of *Hydrogen* is of infinite length. The inflexion point of the last *Helium* neutron process generates the proton of the *Hydrogen*.

The neutron processes of the *Hydrogen* together with the electron process *blue shift* drive and the proton process cover lasts for infinite time.
There is no quantum entropy generation during the *Hydrogen* process. Quantum, generated for the previous periods and the *Quantum Membrane* remains with the only impact of the *Hydrogen* electron process:

$$\lim e_e^H = \frac{dmc_H^2}{dt_{iH}\varepsilon_{xH}}\left(1 - \sqrt{1 - \frac{(c_H - i_H)^2}{c_H^2}}\right) = 0$$

8F1

Space-time of infinite length means decreasing quantum communication.
As there are no more quantum entropy generation and the electron process of the *Hydrogen* is step by step is disappearing as utilised by its neutron process of infinite length, the *Quantum Membrane* is permanently losing of its intensity.

8F2 $\lim e_H^n = \dfrac{dmc_H^2}{dt_H}\sqrt{1-\dfrac{(c_H-i_H)^2}{c_H^2}}\left(\sqrt{1-\dfrac{i_H^2}{c_H^2}}-1\right)=0$;   neutron process intensity

8F3 $\lim e_H^p \dfrac{dmc_H^2}{dt_H}\left(1-\sqrt{1-\dfrac{i_H^2}{c_H^2}}\right)=0$;   proton process intensity

8F4   Without electron process *blue shift* impact, with quantum communication decreasing – quantum gradient is approaching *zero*: $\lim\left|\dfrac{dmc_H^2}{dt_H}\sqrt{1-\dfrac{(c_H-i_H)^2}{c_H^2}}\right|=0$

8F5   The duration of the inflexion point of the *Hydrogen* becomes of its infinite length and of infinite low intensity: $\lim dt_H = \infty$

8F6   The space expansion, with no quantum generation, is *zero*: $\lim ds_H = 0$

8F7   As result of the decreasing overall expansion, the lost energy drive and decreasing quantum communication $\lim c_H = \dfrac{ds_H}{dt_H}=0$

No matter exists without the definition of *time* and *space*.

8F8   Expanded matter with no energy drive is collapsing into *neutron state* with infinite intensity $\lim \dfrac{dm}{dt}=\infty$

8F9   and the new *proton process* starts with infinite intensity and infinite value of quantum communication $\lim c = \infty$

This way the inflexion between the *Hydrogen* and the *First* element of the new cycle will be of *infinite* value of intensity (the *Big Bang*), also with infinite value of the speed of quantum communication. The *First* element starts from infinite value *of space-time,* (opposite to the *Hydrogen* ending with zero *space-time*), this way having only collapse – neutron process.

In the infinite chain of elementary processes the *Last* element (the *Hydrogen*) and the *First* element have quantum entropy gradients and speeds of quantum communication of opposite end values. The table below demonstrates the distance (space) and time components of the speed value of the quantum communication:

8F10

$\lim c = \lim \dfrac{ds}{dt}=0$	$(\cdot)$	$\xrightarrow{\hspace{2cm}} \; \dashrightarrow$	$\lim c = \lim \dfrac{ds}{dt}=\infty$
	$\xrightarrow{\hspace{2cm}} \; \dashrightarrow$	$(\cdot)$	

All these above give specific features to the *Hydrogen*.

While accumulating and with that physically expanding space, *Hydrogen* itself as "quantum generator" does not contribute to the expansion of the *Quantum System of Reference* (quantum space) as *Hydrogen* does not generate *quantum entropy* and *energy quantum*.

Ref. 8F1   8F1 characterises the end stage of the *Hydrogen* process, the status, when no other elementary processes exist anymore. Otherwise *Hydrogen* has its natural electron process of infinite low intensity, acting for infinity.

8.2.2. Vacuum

Vacuuming out elementary processes from the *Quantum Systems of Reference* does not mean at all pumping out the quantum content of the *space-time.*

Elementary processes are the materialised appearances of the quantum world. Neutron process, the re-creation of the mass status cannot happen without proton process cover (the *Strong Interrelation*) and without electron process *blue shift* drive (the *Weak Interrelation*). Therefore, taking mass and energy out of any elementary system is only possible if the *blue shift* drive is also taken. *Blue shift* drive is quantum impact, transferred by the *Quantum System of Reference* (*Quantum Membrane*).

Space-time is continuous and does not exist without *quantum.*
Energy quantum of *Quantum System of References* (*space-times*)
                                                    communicate with each other.
Quantum Membrane with taken out elementary processes is rehabilitating through the boundaries of the *space-time* (of the equipment) otherwise the *space-time* (and also the equipment) is collapsing and as such does not exist anymore.
Equipment, withstanding vacuum impact in our technical environment, all have certain temperature level. There is no way cooling down to absolute *zero* – as proof of the existing *Quantum System of Reference* inside.

Vacuum systems of equipment communicate with the *Quantum Systems of Reference* (external relative to the equipment). Vacuum equipment may be lighted through, as proof light impact is propagating within the internal *Quantum System of Reference* (of systems or equipment under vacuum).

Universe as *space-time* under vacuum is transmitting *blue shift* signals and quantum impacts – as proof of the existing *Quantum System of Reference* of the external *"space".*

8.3
Shortening *space-time* – increasing intensity

If we compress the *space-time* of elements, compressed quantum processes (of elements) will be of increased intensity: The *space-time* of natural elementary processes of elements, including the *blue shift* process and the *Strong Interrelation* are in this case externally and "artificially" changed.
Two components are important here:
(1) the proton-neutron inflexion point with the defined positive and negative   Ref
    gradients of the change and   8D1
(2) the *blue shift* impact and the *blue shift* conflict of the element, driving   8D2
    neutron process with increased intensity, result of the external impact.

(Compression in conventional terms may equally impact gaseous, solid and liquid statuses as well. Relating it to gaseous status is more practical.)

8G1 The intensity of the quantum gradient is:

$$q = \left| \frac{dmc_c^2}{dt_o} \sqrt{1 - \frac{(c_c^2 - i_c^2)}{c_c^2}} \right|$$

Ref. Tab 8.1 c_c establishes the *space-time* of the process; result of the increased intensity of the *blue shift* effect – consequence of the external compressing impact with reference to Table 8.1.

8G2 The element is not changing at all, since elementary characteristics are without change

$$\varepsilon_x = \frac{\varepsilon_p}{\varepsilon_n} \sqrt{1 - \frac{(c - i)^2}{c^2}}$$

but the intensity of the electron process *blue shift* impact, consequence of the change of the *space-time*, is increased:

8G3
$$E_c = \frac{dmc_c^2}{dt_i^c \varepsilon_i^c \varepsilon_x} \left(1 - \sqrt{1 - \frac{(c_c - i_c)^2}{c_c^2}}\right); \quad \text{and} \quad e_c = \frac{dmc_c^2}{dt_i^c \varepsilon_x} \left(1 - \sqrt{1 - \frac{(c_c - i_c)^2}{c_c^2}}\right);$$

(*c* in the index means "compressed")

External energy (compression) results in *blue shift* conflict. This is an intensity increase, with inflexion point and gradients of increased intensity, establishing the intensity level of the following elementary processes, as result of the *Strong* and the *Weak Interrelations*.

If we would have here certain measurement devices, we could measure the change of the quantum communication as well = increased.

Increased quantum communication without space increase means gas status might be turning into liquid status. Otherwise the space increase would give compensation and there would be no intensity increase. The explanation is **Ref** that the intensity of the neutron process is increasing, while the elementary **8G2** proton-neutron process relation, reference to 8G2 stays unchanged.

This is not about condensing. In the contrary it represents much higher energy content and until the external pressure is on, it keeps its liquid format. Gas is condensing, when energy *blue shift* intensity has been taken from it. With the increase of the external *blue shift* impact, external compression energy, the status could be even more developed, reaching solid status.

The reason is the change of the *space-time* of the elementary process: In gaseous status the proton process has its overwhelming intensity dominance against the neutron process. The impact of the generating electron process *blue shift* surplus in normal *space-time* circumstances keeps the element in its natural gaseous state.

If the case only would be about the simple absolute increase of the intensity difference, like in the case of external heating of the gas, the absolute difference between the intensities of the proton and the neutron processes **8G3** would increase: $p - n = \Delta$ and $x \cdot p - x \cdot n = x \cdot \Delta$

And the gas would be even "of more gas status".

The explanation of this still gaseous status is that neither the time system, nor the intensity relations of the *blue shift* process have been changed, but remained at the same value, as the expansion (space increase) of the gas compensates the energy intake by heating.

In the case of compression, the "space" is limited and the generating electron process *blue shift* impact and conflict does not have the chance for correction by expansion. The electron process is driving the neutron collapse with the standard for the element intensity and this way the proton process cover is provided. But the speed of quantum communication of the established new *space-time* changes the status of the element from gaseous to liquid – as the neutron-proton inflexion point becomes of increased intensity, while the element stays the same as it was, still having its standard $\varepsilon_x = const$.

With further increase of the compression the gaseous status could be turned even into solid without in fact changing the original characteristics of the element gaseous in normal dt_i time system and ε_i intensity.

Detailed analysis:

The intensity relation of the proton and neutron processes, made liquid by the compression of the gas corresponds to its standard intensity, in line with the not changing ε_x the intensity coefficient of the electron process of the element. As consequence the element stays the same.

Ref
8G2

Ref
7A1-
7A4

With the increase of the external *blue shift* impact, with reference to Section 7 and 7A1-7A4, the drive of the neutron collapse is higher than that usually is in normal circumstances of the gaseous status.

The increased drive results in inflexion point of increased intensity, established by quantum entropy gradients. The consequences are increased speed of quantum communication and *space-time* of increased intensity. The intensities of the neutron and proton processes obviously become increased, while their relation remains unchanged.

$$\frac{dmc_c^2}{dt_o \varepsilon_x}\sqrt{1-\frac{i_c^2}{c_c^2}}\left(1-\sqrt{1-\frac{(c_c^2-i_c^2)}{c_c^2}}\right) > \frac{dmc^2}{dt_o \varepsilon_x}\sqrt{1-\frac{i^2}{c^2}}\left(1-\sqrt{1-\frac{(c^2-i^2)}{c^2}}\right) \qquad 8H1$$

The *blue shift* drive of the neutron collapse in 8H1 with increased quantum communication is more than in normal circumstances.

In the *space-time*, established by the compression the element, earlier gaseous, becomes of liquid status.

The liquid status in this *space-time* still means *blue shift* conflict, but of less value than that is within the original *space-time* without the compression. And the *blue shift* drive in this *space-time* is more intensive.

There is a certain balance between the proton and the neutron processes, relating to the electron process intensity as being the drive:

8H2   *Proton* process of high and *neutron* process of low intensities mean generation of increased number of electrons, but all with low intensity, causing this way low intensity neutron collapse (as per the initial condition) and electron process *blue shift* surplus, as the gradient of electron process generation is high (in line with the initial condition of the high intensity proton process).

Neutron process of high and *proton* process of low intensities mean generation of less number of electrons, but all with high intensity, causing this way high intensity neutron collapse (as per the initial condition) and electron process *blue shift* deficit, as the gradient of the electron process generation is low (in line with the initial condition of the low intensity proton process).

The *space-time* of the liquid state (result of the compression) means increased neutron process potential to the element. Until the compressed status is provided, this *space-time* is the rule and the element is in liquid status. Once the external *blue shift* impact and with that the compression is released, the *space-time* of the liquid gas becomes expanded.

The increased neutron process potential of the liquid status in conventional terms is of increased energy content. The acting during the compression process external *blue shift* impact is building into the elementary structure. Now, with the "release" of the compression (with the release of the external *blue shift* impact) the *space-time* (in other words the *Quantum Membrane*) of the element in liquid status will be expanding and gaseous with increased electron process *blue shift* surplus.

The two sides of the quantum communication now are replaced:
The element in liquid status is impacting the surrounding environment by its accumulated electron process *blue shift* surplus.

> [If compressed gas is expanding, the expansion takes energy from the vessel which kept the gas pressed. The temperature of the vessel can easily be even much lower than the temperature of the environment, as the quick expansion results in intensity effect. But this is a different case.]

Ref.
S.3.1
S.3.2   With reference to S.3.1, all *Hydro-Carbons* are of this type. With reference to S.3.2, *water* is of different principle.

Elements and elementary compositions in liquid status of certain constant space can be turned into solid. Further increase of the speed of quantum communication needs additional external *blue shift* (energy) intake. Reaching solid state this way means, with reference to Section 10.1 new elementary structure, under infinite high *blue shift* of *Earth's plasma*. This is contrary to freezing as solid status, when *blue shift* surplus is taken away. In this case the *blue shift* surplus is fully utilised.
New elements can only give back the "energy" intake if melted.

Ref.
S.
10.1

8.4
Solid structures with *blue shift* conflict

In the case of minerals of elements or elementary compositions (in solid powder format) the increase of the intensity of the electron process *blue shift* drive (as from external impact) results in the growth of the intensity of the neutron process, strengthening with that the solid status. Further increase results in internal *blue shift* conflict and even in structures of earlier neutron process dominancy the mix of minerals is melting.

Melting is about the conflict between the intensified internal electron process *blue shift* of the neutron process drive and the external *blue shift* impact (possible heat impact). The higher is the conflict, the more liquid is the status of the melted mix.

Once the external impact is over, the conflict is off and the mix is hardening.

The difference between the liquid status with increasing internal intensity (of the previous 8.3 Section) and the liquid status of the mix here is that

- in the case of increased internal intensity, the initial gaseous state was with *blue shift* surplus; the increased intensity of the original surplus is turning the gas into liquid state, intensifying the neutron collapse;
- here the original state is solid, with *blue shift* deficit.

The process can be divided into three parts:

(1) The external *blue shift* impact (possible heat) increases the internal *blue shift* conflict of the solid structure (it is warming up). This conflict through the increase of the intensity of the internal *blue shift* impact intensifies the neutron process. The increasing intensity of the proton process – as consequence of the intensity increase of the inflexion point – results in increasing number of electron processes, with *blue shift* drive of the solid structure of limited intensity (as reference to 8H2).

(2) At a certain stage of the internal *blue shift* conflict caused by the external *blue shift* impact there will be a balance between the intensities of the neutron and proton processes of the original mineral still at quasi solid status (heated up to the melting point).

(3) The liquid state is consequence of the further increase of the internal *blue shift* conflict, consequence of the external electron process impact – responsible for the liquid state.

Ref
S.8.3

Ref
8H2

The solid status have however two stages: (1) the increased intensity of the inflexion point and quasi balance between neutron and proton processes as the state is coming back from liquid to solid; (2) the normal intensity of the inflexion point, with natural characteristics of the element with different proton and neutron process intensities.

As the proton and neutron processes at the increased state are of increased intensity, the solidified mineral gives off *blue shift* surplus for reaching the natural state. This *blue shift* generation and impact means, the solid mineral has *blue shift* surplus. These solid systems of minerals therefore have powder status, as the effecting *blue shift* impact causes conflict.
(Examples: *Coal, Silicon, Sulphur, Calcium*)

S.
8.5

8.5
Extending (lengthening) *space-time* – release of intensity

Learning from the experience of the increasing intensity of the inflexion point, the release of intensity and extension of the *space-time* shall result in generation of *blue shift* surplus – energy generation.
- Any external *blue shift* impact "energy intake" by elementary processes results in the increase of the intensity of the *space-time*. Energy intake in conventional terms means: compression, heating, mixing etc.
- Any *blue shift* release by elementary processes shall mean the extension of the *space-time* with decreasing intensity. *Blue shift* release and energy generation in conventional terms means: cooling, expansion, light etc.

Nature provides minerals as mixture of elements with neutron process and proton process intensity dominances. As proving examples, *oxides, nitrates, carbons, silicates, sulphates, limes* and all *other* mixtures of this kind are in solid status. The natural advantage of these compositions is that there are elements with proton process dominance and electron process *blue shift* surplus incorporated and acting within these minerals. This provides *blue shift* surplus potential (energy) to these natural mixtures.

The intensity of the inflexion point of elementary processes corresponds to specific *space-time*, characteristic of the element and elementary structures. Less intensity value of the inflexion point also means less intensity of the internal electron process *blue shift* impact (as the establishing speed of quantum communication is of less value).

We can formulate this statement in different format in more general:
Our *space-time* and our speed of quantum communication determine the status of elements and elementary compositions within our time system.

Ref
S.5   Impacting the inflexion point of elementary neutron-proton-electron process relations means changing the status of elements.

5A4   With reference to Section 5 and 5A4, there is a significant difference between categories of electron process *blue shift* surplus (as one of the characteristic for the gaseous status) and increased electron process intensity (as one of the features of the solid status). Elements with neutron process intensity dominance (solid structures) have electron process with increased intensity.

The intensity of the inflexion point of an element can only be decreased by releasing on the intensity of the electron process *blue shift* impact – with cooling or decompressing as it is listed in Section 8.3.

Ref S.8.3

Mixing certain minerals with high inflexion point intensity with other elements or minerals with less intensity of the inflexion point (a kind of mixtures listed above), the integrated value of the intensity of the inflexion point of the mix will be between the two initial intensities. The reason is that the proportions of the acting components of the mixtures are changing.

Ref. S.5 Diag. 5.3

With reference to Section 5 and Diagram 5.3, elementary components of mixtures are communicating with each other. Neutron process intensities, coded by *Nature* into elements are different. Elements are streaming to have balanced elementary structures.

Electron process of Element A with electron process *blue shift* surplus will provide *blue shift* impact and proton process cover to Element B with more intensive electron process *blue shift* impact but in relative deficit of it. In this case, the intensity of the inflexion point belonging to the same neutron process earlier part of elementary cycle of Element B, will be belonging to Element A – with inflexion point of less intensity, as the drive of the neutron collapse now becomes more in numbers but of less intensity.
Electron process of Element B with electron process *blue shift* deficit, but of more intensity in parallel will drive the neutron process of Element A and will cover the collapse by its (Element B) proton process.

Elementary compositions of both, A and B this way come closer to an integrated balanced status.
The unique feature of the above communication is that as Element A is with relative electron process surplus and Element B is with deficit, the proportion of the simultaneously acting electron process *blue shift* drives from Element A will be more. Therefore the proportion of the re-creation of Element A within the mixture is more.

(*Oxides, nitrates, carbons, silicates, sulphates, limes* and *others* are acting and communicating internally this way.) Elements with electron process *blue shift* surplus and inflexion point of less intensity will expand their electron process *blue shift* impact more and more.

This is in general in line with the acting rule of *Nature*:
Elementary cycles go in the direction of expansion (energy generation).
The integrated value of the intensities of the inflexion point of elementary processes (the integrated value of the intensities of the *blue shift* of electron processes as indicator) is decreasing. The proof is the existing and increasing elementary dominance of the *Hydrogen. Quantum entropy gradients* are in balance and once the expansion ends, the overall collapse will start.

The increasing number of elements with less inflexion point intensity (and *space-time* with less speed of quantum communication) within the mixture will permanently drive the process into this direction and "generate" *blue shift* surplus, decreasing with that the integrated electron process *blue shift* deficit of the mixture. This generated *blue shift* surplus as benefit however cannot be source of energy generation as it stays within the mix.

8I1   If mixing minerals with increased inflexion point intensity with elements of *O, Ni, C, S, Ca, Si* and *Al, Mg, Cl, K, P, Na, Ni, Fl, Ti* and others, with inflexion point intensity of less value (and with neutron and proton process intensities close to balanced status), the integrated intensity value of the neutron-proton process inflexion point will be decreasing. The generation of *blue shift* surplus and the recreation of elements with inflexion point of less intensity value, in the case of sufficient reserves of the mineral, in fact will be permanent and lasting for long.

For measuring the effect, minerals have to be mixed very well. Quantum communication shall be the best as possible therefore all elements have to communicate with each other at the optimum, as the case requires it, the best possible way. Any *blue shift* conflict impact on the mix, (1) heat effect of the mixing process, (2) heat from the environment (in conventional and global terms = energy intake by the mix) works against the decrease of the intensity of the inflexion point. Intensive cooling therefore helps for speeding up the process and measure the *blue shift* generation.

Ref.   Mixing is not easy as all minerals and added elementary components are in internal balance. Water helps for accelerating and making the mixing more
S.3.2   efficient. With reference to Section 3.2. *Hydrogen* keeps *Oxygen* in the water in liquid status.
> The *blue shift* demand of *Hydrogen* within the water will also be impacting the other elements of the mix. The *Hydrogen* content of the mix will stay constant, but the mix, loosing on the intensity of the inflexion point at integrated level – as *Hydrogen* also utilises *blue shift* impact of elements without *blue shift* surplus and with that releasing inflexion point intensity – becomes easily liquid.
> The *Oxygen* of the water within a mix is with the highest *blue shift* generation potential. Therefore if any of the elements of the mixture provides *blue shift* drive to the *Hydrogen* it replaces an electron process drive from the *Hydrogen-Oxygen* relation and since the *Hydrogen* content is constant within the mix, by that the *Oxygen* cycle is rehabilitating and as gas can leave the mix.

In the case the mix with water is heated or pressed, the external *blue shift* impact intensifies the elementary communication and the *Oxygen* release will be therefore more intensive. At the same time the external *blue shift* intake works against the decrease of the integrated intensity of the inflexion point of the mixture, it increases the intensity.

The intensity of the inflexion point of elementary components or minerals, driving *Hydrogen* neutrons will also be of higher value. As result of the heating the quantum communication of the impacting elements listed in 8I1 within the mix starts from higher inflexion point intensity. And with that the *blue shift* generation potential of the mix is less. (Elements or components will be not changing as ε_x stays unchanged, but ε_i^c becomes of higher, dt_i^c of shorter value.)

Ref 8I1

Cooling down after does not help, as cooling decreases the intensity at integrated level. There would obviously be a certain shift increase in intensities, but this would be significantly less than that it could be without the external *blue shift* intake.

Electron process intensity of elements with inflexion point of less intensity and with *blue shift* surplus will be driving and covering neutron processes of elements with increased intensity but with electron process *blue shift* deficit. This effect in long term results in the increase of the proportions of elements with less intensive inflexion point, with less intensive electron process *blue shift* drive and *blue shift* surplus.

8.5.1. *Intensity release indeed*

S. 8.5.1

An elementary process is taken with inflexion point establishing a solid state with certain intensity. This state belongs to our *space-time* with certain speed value of quantum communication and *blue shift* impact.

$$\frac{dmc^2}{dt_o}\sqrt{1-\frac{(c-i)^2}{c^2}}$$ is the value of the quantum entropy gradient

8J1

If the speed of quantum communication becomes of less value – which in relative terms is identical to the time component of the inflexion becomes longer – the intensity of the inflexion point and this way the new *space-time* of the element becomes of less value.

The mass-energy balance of the element with certain standard electron process intensity value in our time system will have energy and *blue shift* surplus in these new circumstances:

$$\frac{dmc_s^2}{dt_{is}\varepsilon_x}\left(1-\sqrt{1-\frac{(c_s-i_s)^2}{c_s^2}}\right) < \frac{dmc_o^2}{dt_i\varepsilon_x}\left(1-\sqrt{1-\frac{(c-i)^2}{c^2}}\right)$$

8J2

The explanation and the proof of the relation above are:

$$c_s < c; \quad \text{and} \quad \left(1-\sqrt{1-\frac{(c_s-i_s)^2}{c_s^2}}\right) < \left(1-\sqrt{1-\frac{(c-i)^2}{c^2}}\right); \quad \text{and} \quad dt_{is} > dt_i$$

8I3

The existing electron process *blue shift* surplus can be realised by releasing electron process *blue shift*, intensity of

8J4
$$\frac{dmc_s^2}{dt_{is}\varepsilon_x}\left(1-\sqrt{1-\frac{(c_s-i_s)^2}{c_s^2}}\right)$$

As explanation we can state that otherwise the "external" *blue shift* intensifies the inflexion point and increases quantum communication.

Once the intensity of the inflexion has been changed and becomes of less intensity, the less intensity will influence not just the given solid state of the element, but the state of the element will be with electron process *blue shift* surplus. The element in solid state will generate heat.
Either the heating surplus is released or the element will change its state and accommodates itself to the new intensity circumstances.

For slowing down the time system elements they have to be accelerated.

S.
8.6

8.6
Integrating function of the *Quantum Membrane*

Space-time and inflexion point intensities vary – quantum entropy gradients vary.
Certain *space-times* relate to certain specific quantum entropy gradients!

Quantum Membrane is about integrated and consolidated speed of quantum communication and this way integrated intensity of the inflexion point.
While the intensities of the *blue shift* impacts are variant,

8K1
$$e_{x1}=\frac{dmc_1^2}{dt_i\varepsilon_{x1}}\left(1-\sqrt{1-\frac{(c_1-i_1)^2}{c_1^2}}\right);\;\ldots\quad e_{x1n}=\frac{dmc_n^2}{dt_i\varepsilon_{xn}}\left(1-\sqrt{1-\frac{(c_n-i_n)^2}{c_n^2}}\right)$$

the intensities of the inflexion point of elements, whatever are their number and their status will be of identical integrated value.
The intensity of the inflexion of the mix will be belonging to the same speed of quantum communication, as the inflexion point is about neutron and proton processes, the intensities of which have been established by the *Quantum Membrane*:

8K2
$$e_p=\frac{dmc^2}{dt_o}\left(1-\sqrt{1-\frac{i^2}{c^2}}\right);\quad\text{and}\quad e_n=\frac{dmc^2}{dt_o}\sqrt{1-\frac{(c-i)^2}{c^2}}\left(\sqrt{1-\frac{i^2}{c^2}}-1\right);$$

The communication via the inflexion point shall go with the same quantum communication speed. The operating speed of quantum communication of the *Quantum Membrane* is one and the same.
The *Quantum Membrane* has its integrating function: Different elements with different quantum gradient intensities communicate at equal speed of quantum communication.

The integrating effect is result of the *blue shift* conflict of the electron processes of the elements of the mix.

8.7
Intensity coefficient of the electron process

$dt_i = const$, denominator is increased: $\varepsilon_x \uparrow$

$$e_e = \frac{dmc^2}{dt_i \varepsilon_x}\left(1 - \sqrt{1 - \frac{(c-i)^2}{c^2}}\right) = \frac{dmc^2}{dt_i \varepsilon_x}\left(1 - \sqrt{1 - \frac{(a\Delta t)^2}{c^2}}\right)$$

with $a\downarrow \Rightarrow \varepsilon_x \uparrow$ because with the decrease of the value of the acceleration the intensity of the electron process is down.

when $\varepsilon_x \uparrow \Rightarrow \varepsilon_n \downarrow$ as $\varepsilon_n \downarrow$ means: $dt_n \uparrow$ and $\varepsilon_x = f\left(\frac{1}{a}\right)$

		Internal relations		External impact	*power*
$\varepsilon_x \uparrow$	$\varepsilon_n \downarrow$ and $\varepsilon_p \uparrow$	*Unlimited* internal energy power	H	Soft but unlimited energy support	+ + +
	$\varepsilon_p \uparrow$ and $\varepsilon_n = const$	*Increased* internal energy power	O	Strong but limited energy support	+ +
	$\varepsilon_n \downarrow$ and $\varepsilon_p = const$	Internal energy *surplus,* weakening demand	C, S, Si Ca, Cl …	Soft and limited energy support	+
$\varepsilon_x \downarrow$	$\varepsilon_n \uparrow$ and $\varepsilon_p = const$	*Increasing* demand, *weakening* internal energy	Al, Mg, Na, K..	Soft and limited energy demand	–
	$\varepsilon_p \downarrow$ and $\varepsilon_n = const$	*Increased* internal energy demand	Pb, U, Pl, Th…	Strong but limited energy demand	– –
	$\varepsilon_p \downarrow$ and $\varepsilon_n \uparrow$	*Unlimited* energy demand	no	Strong and unlimited energy demand	– – –

Table 8.2

The coefficient of the electron process intensity is:

$$\varepsilon_x = \frac{\varepsilon_p}{\varepsilon_n}\sqrt{1 - \frac{(c-i)^2}{c^2}} = \frac{\varepsilon_p}{\varepsilon_n}\sqrt{1 - \frac{(a\Delta t)^2}{c_2}}\ ; \qquad (a\Delta t = const)$$

The absolute balance is provided, but in intensity terms, the proton process (energy generation) is ahead of the neutron process (workload).

8L3
$$\frac{dmc^2}{dt_o}\left(1-\sqrt{1-\frac{i^2}{c^2}}\right)=\varepsilon_{x\uparrow}\frac{dmc^2}{dt_o}\left(\sqrt{1-\frac{i^2}{c^2}}-1\right);\quad as\;\;\varepsilon_n=\frac{\varepsilon_p}{\varepsilon_x}\sqrt{1-\frac{(c-i)^2}{c^2}}$$

8L4
$$as\;\;\varepsilon_{x\uparrow}>\varepsilon_x:\quad\frac{dmc^2}{dt_p}\left(1-\sqrt{1-\frac{i^2}{c^2}}\right)>\frac{dmc^2}{dt_n}\sqrt{1-\frac{(c-i)^2}{c^2}}\left(\sqrt{1-\frac{i^2}{c^2}}-1\right)$$

This is equivalent to internal energy generation of less intensity and this way less value of electron *blue shift* impact.

8L5
$$\varepsilon_{x-updated}=\frac{\varepsilon_{p\uparrow}}{\varepsilon_{n\downarrow}}\sqrt{1-\frac{(c-i)^2}{c^2}}>\varepsilon_x$$

The consequence is:
- local quantum membrane of the element is of less intensity.
- quantum communication is of less value.

The internal balance has two sides: - increasing ε_x, or - decreasing ε_x
- increasing ε_x means generating electron process *blue shift* surplus,

Ref.
Table
8.2

- decreasing ε_x means developing electron process *blue shift* deficit.

External impact and internal relations given in the variants of Table 8.2.
- generating *blue shift* surplus is providing energy and proton cover,
- developing *blue shift* deficit is taking energy and proton cover from.

S.
8.8

8.8
About *fire* again

With reference to Section 3.3, fire is *blue shift* conflict of infinite intensity.

Ref
3C1

As with reference to 3C1 the increased *blue shift* impact can be written as:

$$X\frac{dmc^2}{dt_i\varepsilon_x}\left(1-\sqrt{1-\frac{(c-i)^2}{c^2}}\right)=X\frac{dn}{dt_i\varepsilon_x}q$$

In the case of acting external *blue shift* impact with limited space and with no expansion possibility for compensating the effect, the increased *blue shift* conflict increases the intensity of the inflexion point and speeds up quantum communication. In fact changes the *space-time* of the element.

The intensity of the inflexion point is establishing the *space-time* in harmony with the intensity of the *blue shift* drive and the load of the *Quantum Membrane* as $i=\lim a\Delta t=c$. Therefore an element within a certain *space-time* has its known standard status.

In the case of changing circumstances the element first tries to keep its standard status: If any additional external *blue shift* impact, with the increase of the intensity, the element tries to modify the circumstances to keep the status.

In this certain case an expansion against *blue shift* impact intensity increase may help.

With reference to 3C2 and 3C3 and 8B7 the situation can be written as:

Ref
3C2
3C3
8B7

$$\frac{dmc^2}{dt_i\sqrt{1-\frac{u^2}{c_s^2}}\varepsilon_x}\left(1-\sqrt{1-\frac{(c_s-i_s)^2}{c_s^2}}\right) = \frac{dn}{dt_{is}\varepsilon_x}q_s \qquad \text{and} \qquad \frac{1}{X} = \sqrt{1-\frac{u^2}{c_s^2}}$$

The integrated *blue shift* impact is of increased intensity as

$$dt_{is} = dt_i\sqrt{1-\frac{u^2}{c_s^2}};$$

8M1

where c_s is taken as the speed of quantum communication related to the *Quantum Membrane* of the fire of increased intensity. $\lim u = c_s$

is the condition of the fire,

The same way as the *blue shift* conflict is: $X = f\left(\dfrac{u}{c_s}\right)$

$$\frac{dmc_s^2}{dt_{is}\varepsilon_x}\left(1-\sqrt{1-\frac{(c_s-i_s)^2}{c_s^2}}\right) = \frac{dn}{dt_{is}\varepsilon_x}\frac{dmc_s^2}{dt_o}\sqrt{1-\frac{(c_s-i_s)^2}{c_s^2}}$$

8M2

The main points here are the change of the speed of quantum communication and the value of the quantum entropy gradient:

Ref
8J1
8M3

With reference to 8I1 it will be: $q_s = \left|\dfrac{dmc_s^2}{dt_o}\sqrt{1-\dfrac{(c_s^2-i_s^2)}{c_s^2}}\right|$

The time system of the electron process is consequence of the *space-time.*

In the case of fire, the *blue shift* impact of the element in fire is of slowed down time flow and of infinite intensity. The quantum gradient the same way is of infinite intensity (frequency) and increased to infinity quantum communication speed.

Quantum entropy gradient (as drive of the quantum space in the inflexion point) of infinite intensity cannot contribute to proton process generation, since the full cycle would be with processes and quantum communication of infinite intensity. In this case the element will be destroyed, as the neutron collapse is of infinite intensity, driven by electron process *blue shift* also of infinite intensity.

The quantum entropy gradient is impacting the *Quantum Membrane*, as quantum does not take and accumulate energy.

The *Quantum Membrane* transfers the impact, which is certain well known in conventional terms impulse: *light*!

Elementary process is destroyed with all its process components, the *Strong* and *Weak Interrelations*. The result is *blue shift* impact of infinite (as it was) frequency, causing *blue shift* conflict, in conventional terms: *heat*.

9
Inflexion point

Changing *space-time* means the intensity change of the generation of the energy quantum. Changing space-time is equivalent to the change of the speed of quantum communication. The speed of quantum communication establishes the intensity of the inflexion point!

Inflexion point is not equal to status of rest or to the time system of rest, as those as such do not exist. Inflexion point is with the certain feature of zero motion status, the certain common basis for comparing events.
Space and the *Quantum System of Reference* are one and the same.

$$c = \frac{ds}{dt}$$ The speed of quantum communication (the propagation of the quantum impact) is equivalent to *the intensity of the extension of the space = result of the generation of energy quantum.*

9A1

$$qe_o = \frac{dmc^2}{dt_o}\sqrt{1-\frac{(c-i)^2}{c^2}}$$ The intensity of the quantum entropy gradient acting at the inflexion point keeps the *Quantum System of Reference* at basic load.

9A2

Quantum entropy has its format at the moment of its generation as well, expression of $$qe_i = \frac{dmc^2}{dt_o}\sqrt{1-\frac{i^2}{c^2}}\sqrt{1-\frac{(c-i)^2}{c^2}}$$

9A3

c the speed of quantum communication is the *key* for both: the intensity of the inflexion point and the basic load of the *Quantum Membrane.*

dt_o the inflexion belongs to both sides (as end and start) of the proton-neutron and anti-neutron – anti-proton processes.

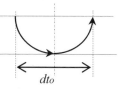

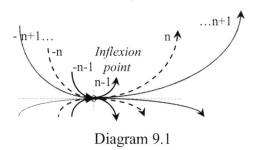

Diagram 9.1

All but one elementary process go through inflexion point. All proton-electron-neutron processes and anti-processes are of different intensities. The intensity of the inflexion point characterises the element.

Diag. 9.1

The *last* element (*Hydrogen*) does not have inflexion point. The *first* element therefore has overwhelming collapse and proton expansion, both of infinite intensity. The collapse of *Hydrogen* neutrons is approaching the inflexion point from both sides (process and anti-process directions).

The load of the *Quantum Membrane* and the intensity of the inflexion point are established by the electron process, the drive of the neutron collapse.

9A4   The decreasing value of the acceleration is compensated by the time component:
$$e_x = \frac{dmc^2}{dt_i \varepsilon_x}\left(1 - \sqrt{1 - \frac{(c-i)^2}{c^2}}\right)$$
$$\lim(c-i) = \lim a\Delta t = 0 = const$$

9A5   while the intensity coefficient of the electron process is constant, characteristic of the element:
$$\varepsilon_x = \frac{\varepsilon_p}{\varepsilon_n}\sqrt{1 - \frac{(c-i)^2}{c^2}}$$

Quantum communication is only "functions" between energy quantum!

The speed of quantum communication, establishing the intensity of the generation of the energy quantum, the intensity of the *inflexion point*, determines the local *space-time* as well.

In our natural circumstances on the surface of *Earth* the speed of quantum communication is $c \approx 300000 \; km/sec$.

The proton function is sphere symmetrical expanding acceleration.

9B1

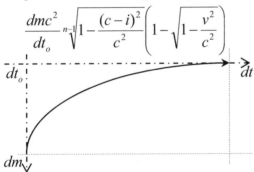

$$\frac{dmc^2}{dt_o} \sqrt[n-1]{1 - \frac{(c-i)^2}{c^2}}\left(1 - \sqrt{1 - \frac{v^2}{c^2}}\right)$$

While the curve characterises the transformation of mass into energy, first of all it demonstrates the change of the time system, as function of the speed of the acceleration, from zero up to $i = \lim v = c$.

dt_o the time system of the inflexion belongs to $v = 0$.

Diag. 9.2   Diagram 9.2

The neutron function is sphere symmetrical accelerating collapse.

9B2

$$\frac{dmc^2}{dt_o} \sqrt{1 - \frac{i^2}{c^2}} \sqrt[n]{1 - \frac{(c-i)^2}{c^2}}\left(1 - \frac{1}{\sqrt{1 - \frac{(i-v)^2}{c^2}}}\right)$$

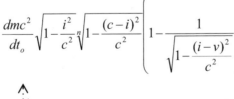

if the basis taken as of dt_i

The character of the neutron process is different. The reason is that the starting mass gradient belongs to $i = \lim v = c$, which for a long time compensates the slowdown and the mass growth of the process. In our *Earth* system of reference until $v = 10^{-6} m/s$ absolute speed of the collapse, mass remains close to the $i = \lim v = c$ speed level.

Diag. 9.3   Diagram.9.3

But the character of the change in fact depends on the time system of the assessment. If time system $i = \lim v = c$ is taken for the basis, the neutron process is similar to the proton process in Diagram 9.2, with the difference that the mass growth at the inflexion point is of infinite high value.

The time systems of the proton expansion and the neutron collapse are permanently changing, but in opposite gradients and opposite directions.

The electron process, the sphere symmetrical expanding acceleration happens at constant $i = \lim v = c$ speed for infinite time.

$$\frac{\frac{dmc^2}{dt_o \varepsilon_x} \sqrt{1 - \frac{i^2}{c^2}} \left(1 - \sqrt{1 - \frac{(c-i)^2}{c^2}}\right)}{dt_i}$$

Diagram 9.4

The question is, what is the capacity, which drives the electron process, if in the proton process almost all mass potential has already been used?

9B3

Diag. 9.4

Electron process continues for the count of mass change dm until the capacity of the *blue shift* is capable to impact the neutron collapse. The continuity of the electron process has been guaranteed by the neutron and proton processes themselves. This is *regulated* by ε_x the coefficient of the intensity of the electron process, which is function of the intensity relation of the proton and neutron processes, while the acceleration is decreasing:

$$\lim(c - i) = \lim a\Delta t = 0 = const \text{ establishes the constant speed.}$$

The distinguishing difference between proton and electron processes is, that once the proton process reaches $i = \lim v = c$, the transformation as such ends, as the process becomes conflicting with the *Quantum Membrane*.

Impact on and conflict with the *Quantum Membrane* is *electron* function.

The question is: What is between the end of the proton process and the birth (generation) of quantum entropy?
The answer is coming from the process itself:

If the end status of the proton process is: $\dfrac{dmc^2}{dt_o \varepsilon_x} \sqrt{1 - \dfrac{i^2}{c^2}}$ and the

9C1

value of the quantum entropy is: $\dfrac{dmc^2}{dt_o \varepsilon_x} \sqrt{1 - \dfrac{i^2}{c^2}} \sqrt{1 - \dfrac{(c-i)^2}{c^2}}$, the process

9C2
Ref
9B3

between the two is the electron process indeed as described above in 9B3:

Electron process is acceleration at constant speed,
blue shift energy impact to *Quantum Membrane*.

The intensity of the electron process is constant, function of the intensity relation of the proton and neutron processes. ε_x the coefficient of the intensity of the electron process and the actual proton (and electron) number keeps the balance of the proton expansion and the neutron collapse.

With reference to Section 6, proton-neutron-anti-proton-anti-neutron cycle has typical configuration, established by the intensities of the processes.

The question is: How can an electron process of constant dt_i time system do drive neutron process of changing intensity?

Electron process *blue shift* drive of infinite time system loads the *Quantum Membrane*. The permanent *blue shift* impact, of specific function for the element results in accelerating collapse. Neutron collapse is the taking in of the energy release of the proton expansion as mass impact. It happens the same way for the anti-processes as well, just the anti-proton expansion is with mass release and the anti-neutron collapse is with energy intake.

The intensities of the release and receipt correlate with the time systems.
The highest intensity of the collapse belongs to the starting (post dt_o inflexion point) proton expansion stage. Less intensive proton and neutron process stages belong to the other end, the time system of dt_i. Energy/mass impacts are acting with the intensity of the expansion and the collapse.

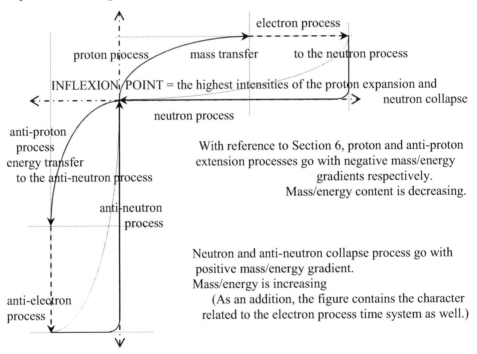

Diag.
9.5

Diagram 9.5

The mass-energy content of the cycle is constant in every its point.

$$dmc^2 \sqrt[n]{1-\frac{(c-i)^2}{c^2}} + \frac{dmc^2}{dt_p \varepsilon_p} \sqrt[n]{1-\frac{(c-i)^2}{c^2}} \left(1-\sqrt{1-\frac{v^2}{c^2}}\right) + \frac{dmc^2}{dt_n \varepsilon_n} \sqrt[n]{1-\frac{(c-i)^2}{c^2}} \left(\sqrt{1-\frac{v^2}{c^2}}-1\right) =$$

9D1

$$= dmc^2 \sqrt[n]{1-\frac{(c-i)^2}{c^2}} + Entropy$$

anti-neutron process (previous cycle) ends – proton process starts:	proton process ends – anti-electron process starts:	
$$\frac{dmc^2}{dt_o} {}^{n-1}\!\sqrt{1-\frac{(c-i)^2}{c^2}}$$	$$\frac{dmc^2}{dt_o} {}^{n-1}\!\sqrt{1-\frac{(c-i)^2}{c^2}}\sqrt{1-\frac{i^2}{c^2}}$$	9D2 9D3
electron process ends – neutron process starts:	neutron process ends – anti-proton process starts:	
$$\frac{dmc^2}{dt_o}\sqrt{1-\frac{i^2}{c^2}}\,{}^{n}\!\sqrt{1-\frac{(c-i)^2}{c^2}}$$	$$\frac{dmc^2}{dt_o}\,{}^{n}\!\sqrt{1-\frac{(c-i)^2}{c^2}}$$	9D4 9D5

and the anti-cycle continues the same way, as per Section 6.

Elementary processes with reference to Section 6 happen in two dimension:
   (1) with mass transfer (as "the normal") process chain; and
   (2) with energy transfer (as opposite to "the normal") less obvious and measurable processes.

As a result of the different load of the quantum entropy gradient, there might be variety of speed values of quantum communication impacting the *Quantum Membrane.* (With reference to 9A1 the speed of quantum communication is identical to speed value of space expansion.)

Energy quantum is communicating and the acting speed within the *Quantum Membrane* is being established as the balance of all impacts.
The acting "energy" of the *Quantum System of Reference* and the *Quantum Membrane* is originating from the *inflexion point.* Quantum gradients, with reference to 9B1 are the heart of the quantum system, gifting the permanent load of quantum communication.

Ref. 9B1

Increase of the intensity of the inflexion point means: increased intensity of the electron process, the proton, neutron processes (and anti-processes) as well. The intensity increase of the inflexion point also means the intensity increase of the local *space-time* of the element.

The standard absolute formula
$$\frac{dmc^2}{dt_o\varepsilon_p}\left(1-\sqrt{1-\frac{i^2}{c^2}}\right)=\frac{dmc^2}{dt_o\varepsilon_n}\xi\sqrt{1-\frac{(c-i)^2}{c^2}}\left(\sqrt{1-\frac{i^2}{c^2}}-1\right)$$
9E1

In intensity terms
$$\frac{dmc^2}{dt_o}\left(1-\sqrt{1-\frac{i^2}{c^2}}\right)\neq\frac{dmc^2}{dt_o}\sqrt{1-\frac{(c-i)^2}{c^2}}\left(\sqrt{1-\frac{i^2}{c^2}}-1\right)$$
9E2

With the increase of the speed of quantum communication the difference between the intensities of the proton and neutron processes is also growing.

The intensity of the inflexion point is:
$$\frac{dmc_x^2}{dt_o}\sqrt{1-\frac{(c-i)^2}{c}}$$
9E3

Mass is result of the permanent transformation of the developing *energy* of the proton expansion into neutron collapse.

Energy is result of the permanent transformation of anti-proton expansion into anti-neutron collapse.

Neutron process is approaching the inflexion point with full *mass*.

Anti-neutron is approaching the inflexion point with full *energy*.

Energy quantum represents both: processes and anti-processes as well.

In the case of *Hydrogen* the intensities of the formulating full mass and full energy statuses are so low that they are not measurable. The inflexion point has not been reached. In the case of the *first element*, during the *Big Bang*, all mass and energy statuses are going through the inflexion point at once.

Ref
S.6

Inflexion point "flower" with reference to Section 6 - with no comments:

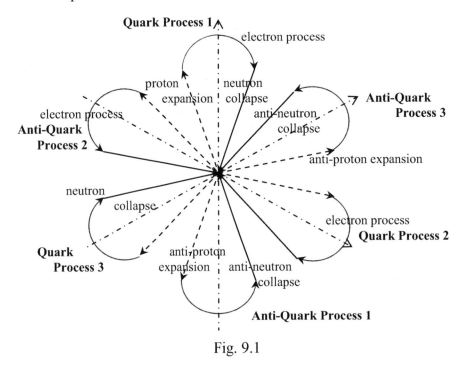

Fig.
9.1

Fig. 9.1

S.
9.1

9.1

Quantum communication at the *inflexion point*

The intensity coefficient of the electron process depends on the relation of the intensities of the proton and neutron processes. The *less* is the value of the coefficient, the more are the *intensities* of the electron process drive, of the neutron process, of the collapse and of the inflexion point.

9F1
$$e_x = \frac{dmc^2}{dt_i \varepsilon_x}\left(1 - \sqrt{1 - \frac{(c-i)^2}{c^2}}\right); \quad \text{as} \quad \varepsilon_x = \frac{\varepsilon_p}{\varepsilon_n}\sqrt{1 - \frac{(c-i)^2}{c^2}}$$

9F2
$$\frac{dmc^2}{dt_p \varepsilon_p}\left(1 - \sqrt{1 - \frac{v^2}{c^2}}\right) = \frac{dmc^2}{dt_n \varepsilon_n}\xi\sqrt{1 - \frac{(c-i)^2}{c^2}}\left(\frac{1}{\sqrt{1 - (u^2/c^2)}} - 1\right); \quad \text{and} \quad u = i - v$$

Elements differ and the speed value of their internal elementary quantum communication – alongside the intensity coefficient of the electron process – is the other distinguishing parameter!

This directly means: c is one of the keys of elementary characteristics!
The speed value of quantum communication is establishing the intensity of the electron process. The intensity of the drive of the neutron collapse, the intensity relation of the proton and neutron processes have been settled accordingly.

The measured weight of elements depends on the intensity of the inflexion point: The highest intensity impact of the neutron collapse is acting at the inflexion point.

The consequences of this finding are of significant importance:
1. The speed value of the quantum communication gives answer to the question what is the normal elementary status of elements at the surface of *Earth*.
    The (only) *eight* elements with proton process intensity dominance and electron process *blue shift* surplus in our natural environment (with $Z < 1$ or $\varepsilon_x > 1$ – the meaning of both are the same as $Z = 1/\varepsilon_x$) are:

    Hydrogen, Oxygen, Nitrogen, Helium,   9F4
    Carbon, Sulphur, Calcium, Silicon   9F5

    can be distinguished relative to their speed of quantum communication:
    The value of c_x
    of the first *four* (with reference to 9F4) is less than the speed of the quantum communication of the *Earth*: $c_x < 300,000$ km/sec; these are of gaseous status;
    of the last four (with reference to 9F5) is more; these are of solid (but powder) status.
2. The speed values of quantum communication of all other elements of the periodic table are more than the speed "of light" measured on the surface of the *Earth*.
    All of these elements have electron *blue shift* deficit and neutron process dominance – but increased c_x speed.

    The higher is the electron process *blue shift* deficit, the more intensive is the drive and more intensive is the neutron collapse, the inflexion point and higher the speed of quantum communication.

Elements in chemical reactions utilise their electron process *blue shift* surplus. Relations establish natural balance: Proton dominance is compensated by neutron dominance.

Elements with speed of quantum communication higher than the "speed of light" on the *Earth* and with slight proton process intensity deficit prove intensive elementary communication. *Magnesium, Potassium (K), Phosphorus, Chlorine, Sodium (Na), Nickel, Fluorine, Titan* etc.

While all these elements are also with slight electron *blue shift* deficit and neutron process dominance, the activity and the energy they provide is significant and very important not just in chemistry but equally to life function as well. The activity and energy provision support of these elements is also explained by the fact that they have higher speed of quantum communication, a certain energy potential and reserve for use.

Other elements with higher atomic weight distinguish themselves by increased speed value of quantum communication.

All elements are for us to use their benefit.
The only criterion is to utilise the given natural potential is without damaging the elementary structure.
-   elements with *blue shift* surplus provide energy directly;
-   elements with *blue shift* deficit help by using their quantum communication potential.

S.
9.2

9.2
Mineral and laboratory status of elements

Conventional physic calls elements *monatomic*, when they cannot keep themselves to their definite atomic structure.
Monatomic is a certain transition status of elements, as the natural speed value of quantum communication of the majority of elements is higher than the quantum communication of the *System of Reference* of the *Earth*.
The intensity of the electron process *blue shift* impact of the element

9G1
$$\frac{dmc_x^2}{dt_i^x \varepsilon_x}\left(1-\sqrt{1-\frac{(c_x-i_x)^2}{c_x^2}}\right) > \frac{dmc_E^2}{dt_i^E \varepsilon_x}\left(1-\sqrt{1-\frac{(c_E-i_E)^2}{c_E^2}}\right); \quad c_x > c_E$$

is with increased speed of quantum communication, above the speed of quantum communication of the *Earth*.

The operating internal elementary *blue shift* impact of *monatomic* elements is of increased value without any external "support". The increased internal *blue shift* impact power and quantum speed keeps elementary components in conflict at micro-level, in a kind of (micro)-powder status. The increased speed of quantum communication makes elementary identification difficult.

With reference to David Hudson's experiments and findings there are proofs of the status. It relates mainly to elements like *Cobalt, Nickel, Cuprum, Ruthenium, Rhodium, Palladium, Silver, Osmium, Iridium, Platinum, Gold, Mercury.* There is however no certain element related significance. All elements might produce similar symptoms.

The symptoms are easy and simple: energy surplus and energy provision.
The energy impact at the inflexion point is of significant power.
Electron process *blue shift* impact with increased speed of quantum communication provides additional potential in any energy relations and conflicts – with all its benefits.

Because of the natural but increased electron process *blue shift* impact, the measured weight of monatomic elements in their mineral stage is higher than that of the same elements in industrial-laboratory conditions.

9.3

What does the

speed of quantum communication of *Earth* gravitation mean?

The sphere symmetrical expanding acceleration of *Earth* with constant speed of $i = \lim a\Delta t = c$ means an exact speed value c. It is measured on the surface of our planet as the *speed of light*
and equal to an *300,000 km/sec* approximate value.

Our known periodic system is not containing all elementary components of the infinite cycle of proton-electron-neutron process chain.
Diagram 9.6 demonstrates the full cycle of elementary transformations: from elements with infinite speed of quantum communication and with infinite value of electron process intensity.

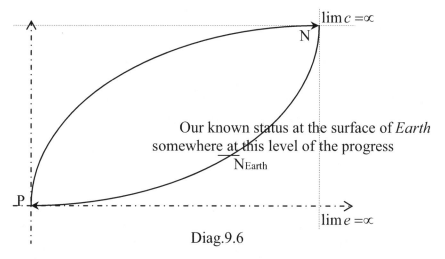

Diag.9.6

P corresponds to total proton process status with only *Hydrogen* content of the *Quantum System of Reference* and with decreasing intensity of the *Quantum Membrane* and speed of quantum communication, approaching *zero*.
N corresponds to the opposite, the total neutron process status, with certain specific elementary content with intensity of the *Quantum Membrane* and the speed of quantum communication, both approaching *infinity*.

Our current status on the surface of the *Earth* is very likely corresponding to a certain N_Earth point, containing elements with neutron and proton processes on the section of the line between N_Earth and P.

(Curve P represents the chain of elementary statuses with decreasing proton process intensity as the Periodic Table also demonstrates it.)

For the calculation of the speed of quantum communication of elements, the starting point is the intensity impact of the electron process:

9H1
$$e_e = \frac{dmc^2}{dt_i \varepsilon_x}\left(1 - \sqrt{1 - \frac{(c-i)^2}{c^2}}\right)$$

Ref.
S.9.2
9F4
9F5

The comparison of the intensity coefficients of electron processes of all elements – in accordance with the arguments in Section 9.2 and reference to 9F4 and 9F5 – show, the speed value of quantum communication of *Earth* shall be between the *c* values of *Carbon* and *Nitrogen*.

Both elements are with electron process *blue shift* surplus, but *Carbon* is solid in its minerals and *Nitrogen* is gaseous.

Gaseous status proves elements have less natural value of quantum communication than *Earth*. Having increased speed value, their conflicting electron process results in gaseous status. While others, also with electron process *blue shift* surplus are in solid (powder) state.

Carbon is taken having speed of quantum communication equal to the one of the *Earth's gravitation.*

9H2   For the simplicity of the calculation, it is taken that $dm = 1$, $dt_i = 1$ and $\left(1 - \sqrt{1 - \frac{(c-i)^2}{c^2}}\right) \cong 1$

as they are equal and quasi equal for all elements.

9H3   Establishing the relativistic value of e_e for *Carbon* $e_e = \frac{c^2}{\varepsilon_x}$ The quantum speed of all other elements is calculated as function of ε_x relative to identical (to the *Carbon*) $e_e \rightarrow c_x = \sqrt{e_e \varepsilon_x}$

- $\varepsilon = 1/Z$ the intensity coefficient of the electron process, the drive of the neutron collapse;

- and $c = 300000$ km/sec, the "speed of light" measured on the surface of the *Earth* as the speed of quantum communication of the *Carbon* element, as with reference to 9F5 *Carbon* is the last element in solid status with electron process surplus and this way the closest to the "speed of light".

Ref.
Book2

With reference to the intensity and elementary data presented in the book of *Quantum Energy and Mass Balance* (Trafford 2009), the results of the calculation are presented in following Table 9.1.

Element	PN	Measured atomic weight	$Z_x^{max} = \dfrac{1}{\varepsilon_x^{max}}$	c_x	comment
Hydrogen	1	1.0079	0.000081	**2718**	approaching zero
Helium	2	4.0026	0.9863	*299924*	
Lithium	3	6.94	1.2961	**343815**	
Beryllium	4	9.012	1.2362	**335777**	
Boron	5	10.81	1.1458	**323266**	
Carbon	6	12.011	0.9868	*300000*	taken as basis
Nitrogen	7	14.0067	0.9860	*298878*	
Oxygen	8	15.999	0.9849	*299711*	
Fluorine	9	18.9984	1.0951	**316033**	
Neon	10	20.17	1.0019	**302286**	
Sodium	11	22.989	1.0743	**313018**	
Magnesium	12	24.305	1.0102	**303536**	
Aluminium	13	26.9815	1.0560	**310340**	
Silicon	14	28.0855	0.9911	*300652*	
Phosphorus	15	30.9737	1.0495	**309348**	
Sulphur	16	32.06	0.9887	*300288*	
Chlorine	17	35.453	1.0699	**312376**	
Argon	18	39.948	1.2028	**331210**	
Potassium	19	39.098	1.0424	**308335**	
Calcium	20	40.08	0.9890	*300334*	
Scandium	21	44.9559	1.1248	**320290**	
Titanium	22	47.9	1.1610	**325403**	
Vanadium	23	50.9415	1.1983	**330589**	
Chromium	24	51.996	1.1503	**323901**	
Manganese	25	54.938	1.1811	**310340**	
Iron	26	55.847	1.1319	**321300**	
Cobalt	27	58.9332	1.1664	**326159**	
Nickel	28	58.71	1.0811	**314007**	
Cuprum	29	63.54	1.1746	**327304**	
Zinc	30	65.38	1.1631	**325698**	
Gallium	31	69.735	1.2327	**335301**	
Germanium	32	72.59	1.2515	**337848**	
Arsenic	33	74.9216	1.2534	**338105**	
Selenium	34	78.96	1.305	**344994**	
Bromine	35	79.904	1.2659	**339786**	
Krypton	36	83.8	1.3104	**345707**	
Rubidium	37	85.4658	1.2927	**340364**	
Strontium	38	87.62	1.2886	**342819**	
Yttrium	39	88.9059	1.2626	**339343**	
Zirconium	40	91.22	1.2634	**339451**	
Niobium	41	92.9064	1.2491	**337524**	

Molybdenum	42	95.94	1.2672	**339961**	
Technetium	43	98.962	1.2843	**342247**	
Ruthenium	44	101.07	1.2799	**341660**	
Rhodium	45	102.9055	1.2697	**340296**	
Palladium	46	106.4	1.2958	**343776**	
Silver	47	107.868	1.2779	**341393**	
Cadmium	48	112.41	1.3244	**347549**	
Indium	49	114.82	1.3258	**347732**	
Tin	50	118.69	1.3561	**351683**	
Antimony	51	121.75	1.3695	**353417**	
Tellurium	52	127.6	1.4356	**361845**	
Iodine	53	126.9045	1.3766	**354332**	
Xenon	54	131.3	1.4134	**359037**	
Caesium	55	132.9054	1.3985	**357139**	
Barium	56	137.33	1.4341	**361656**	
Lanthanum	57	138.9055	1.4188	**359722**	
Cerium	58	140.12	1.3979	**357062**	
Praseodymium	59	140.9077	1.3705	**353546**	
Neodymium	60	144.24	1.381	**354898**	
Promethium	61	*145*	1.3593	**352098**	
Samarium	62	150.4	1.4077	**358312**	
Europium	63	151.96	1.3942	**356590**	
Gadolinium	64	157.25	1.4387	**362236**	
Terbium	65	158.9254	1.4268	**360735**	
Dysprosium	66	162.5	1.4438	**362877**	
Holmium	67	164.9304	1.4433	**362818**	
Erbium	68	167.26	1.4414	**362575**	
Thulium	69	168.9342	1.4301	**361151**	
Ytterbium	70	173.04	1.4536	**364107**	
Lutetium	71	174.967	1.446	**363154**	
Hafnium	72	178.49	1.4606	**364982**	
Tantalum	73	180.9478	1.4603	**364945**	
Tungsten	74	183.85	1.466	**365656**	
Rhenium	75	186.207	1.4643	**365444**	
Osmium	76	190.2	1.484	**367894**	
Iridium	77	192.22	1.4778	**367125**	
Platinum	78	195.09	1.4826	**367721**	
Gold	79	196.9665	1.4747	**366740**	
Mercury	80	200.59	1.4887	**368476**	
Thallium	81	204.37	1.5043	**370402**	
Lead	82	207.8	1.5153	**371754**	
Bismuth	83	208.9804	1.4991	**369761**	
Polonium	84	*209*	1.4695	366093	data are not reliable
Astatine	85	*210*	1.4522	363931	
Radon	86	*222*	1.5622	377463	

Francium	87	*223*	1.5417	374978	
Radium	88	226.0254	1.5494	**375914**	
Actinium	89	*227*	1.5316	373748	not reliable
Thorium	90	232.0381	1.559	**377076**	
Protactinium	91	231.0359	1.52	**372330**	
Uranium	92	238.029	1.568	**378163**	**most intensive**
Neptunium	93	237.0482	1.53	**373553**	
Plutonium	94	*244*	1.5765	***379187***	
Americium	95	*243*	1.5389	374638	
Curium	96	*247*	1.5538	376447	
Berkelium	97	*247*	1.5275	373247	
Californium	98	*251*	1.5422	375039	
Einsteinium	99	*254*	1.5466	375574	
Fermium	100	*257*	1.5509	376095	
Mendelevium	101	*258*	1.5355	374224	not reliable data as the atomic weight forecast is not precise
Nobelium	102	*259*	1.5293	373467	
Lawrencium	103	*256*	1.467	365781	
Rutherforium	104	*261*	1.4909	368749	
Dubnium	105	*262*	1.4767	366988	
Seaborgium	106	*263*	1.4627	365245	
Bohrium	107	*262*	1.4304	361189	
Hassium	108	*265*	1.4354	361820	
Meitnerium	109	*266*	1.4222	360153	
Darmstadtium	110	*271*	1.4453	363066	
Roentgenium	111	*272*	1.4322	361417	

Table 9.1

Table 9.1

Increased speed of quantum communication means increased intensity of the *Quantum Membrane* of the element.

Quantum Membranes are communicating as the time systems of electron processes are quasi equal.

Elements with increased speed of quantum communication are very keen to take external electron process *blue shift* impacts from external *Quantum Membrane* even without proton process cover.

Increased neutron process dominancy makes elements less stable as the deviation between the intensities of the proton and neutron processes makes these elements "soft".

This "softness" is different than that of the elementary status close to liquid state, which is from electron process surplus. The reason of the softness here is the significant difference between the intensities of the proton and neutron processes, the increased intensity of the neutron process.

The large intensity difference does not allow "hard" communication.

As result of the taken external *blue shift* impact the integrated speed value of the quantum communication of the quantum systems is decreasing. This way the value of the quantum speed of the neutron process dominant element is getting less. The elementary structure of the still neutron process dominant element is "hardening" as the intensity of the electron process drive, with stable and constant proton/neutron intensity relations is decreasing.

(In other words: Neutron process dominancy is "eating off" external *blue shift* impacts while proton processes go still the same way behind, because of their less intensity.)

9H4
$$n_{xx}\frac{dmc_{xx}^2}{dt_i}\left(1-\sqrt{1-\frac{(c_{xx}-i_{xx})^2}{c_{xx}^2}}\right)+n\frac{dmc^2}{dt_i}\left(1-\sqrt{1-\frac{(c-i)^2}{c^2}}\right)=$$

$$\text{and } c_{xx}>c_x>c \qquad\qquad =n_x\frac{dmc_x^2}{dt_i}\left(1-\sqrt{1-\frac{(c_x-i_x)^2}{c_x^2}}\right)$$

The decreasing quantum speed is consequence of the common *Quantum Membrane* with speed of integrated quantum communication of less value.

Melting of elements as from minerals results in reduction of the speed of quantum communication, as the melting itself happens in *Earth's* system of reference with gravitation (in sphere symmetrical expanding acceleration with speed of $i = \lim 300000$ km/sec).

Melting or other heat treatments are processes of *blue shift* conflict.
In this case the resolution of the conflict (the hardening of the element) is also about the integration of the element into the system of reference with the given standard speed of quantum communication. The element loses on its total energy/mass potential. The more is the atomic weight, the more is the loss.

9H5

In the case of *Iron*, with natural speed of quantum communication of 321300 km/sec, the loss is 13%.
In the case of *Gold*, the loss is 33%!
In the case of *Uranium*, the elements with most intensive neutron process, the loss is 37%.

as the loss de facto is
$$loss = \frac{c_x^2 - c^2}{c_x^2}$$

Elements with increased speed of quantum communication keep their original characteristic speed value in natural minerals. Minerals with elements of increased speed of quantum communication originate as from the source of the *gravitation* of *Earth*.

Water appears in ice and steam statuses. Its appearance depends on the impacting circumstances.
The components of water are with speed of quantum communication, naturally less than the speed of the quantum communication of *Earth* gravitation.

Therefore de facto, gravitation is the energy supply of the water.

Gravitation keeps water in its water status. Temperature increase or increased *blue shift* conflict turns water into steam, cooling and less conflict brings it back.

With the weakening of gravitation, as with the weakening of the intensity of the *blue shift* conflict steam status in clouds becomes liquid in form of rain.

The *plasma* status within the depths of the *Earth* is *blue shift* conflict of infinite frequency and infinite speed value of quantum communication. The core however as subject to the intensity of the release of the *blue shift* potential of the *plasma* (equal to cooling) has a variety of the resulting characteristics: elements – with different speed values of quantum communication, with different electron, neutron and proton process intensities, as the natural cooling process (decrease of the conflict) is establishing.

Natural minerals of elements with original elementary structure have been acting at the surface of *Earth* with increased speed of quantum communication, as long as this feature has not been impacted.

The usual impacts are the followings:

(1) Melting is the increase of the intensity of the electron process *blue shift* impact of the element by heating.

Solidification of the melted stage goes with the speed of quantum communication, corresponding to the speed of "light" on the surface of the *Earth*.

(2) Dissolving minerals in water results in *blue shift* conflict between water and minerals with increased quantum communication. There is no way two values of quantum communication would act simultaneously.

As result the higher speed value disappears in form of heat generation or without any specific note. The establishing new speed value will correspond to the quantum communication of water (of the *Earth*).

Gravitation is sphere symmetrical expanding acceleration at constant speed establishing by that the certain speed value of quantum communication, corresponding to the "speed of light" on the surface of the *Earth*.

Earth gravitation keeps *Helium, Oxygen* and *Nitrogen* in gaseous status. *Hydrogen* would be in any circumstances gaseous.

Life related vital importance of the *Oxygen,* the *Hydrogen* and the *Nitrogen* is direct consequence of *gravity*.

S.
10

10
Quantum effect of *Earth's gravitation*

The sphere symmetrical expanding acceleration of the electron process is an impact against the *Quantum Membrane*, established by the element itself. Proton process is sphere symmetrical expanding acceleration with the change of the mass/energy status of the proton and most importantly with the constant change of the timeframe of the expansion:

10A1
$$\frac{dmc^2}{dt_o}\sqrt{1-\frac{v^2}{c^2}}\left(1-\sqrt{1-\frac{v^2}{c^2}}\right); \quad \text{up to } dt_i = \frac{dt_o}{\sqrt{1-(i^2/c^2)}}.$$

Proton process continues as electron process: acceleration at constant speed $i = \lim a\Delta t = c$, *work* against the *Quantum Membrane*.

Proton process is an expansion as well, but not against the *Quantum Membrane*. Proton process is the mass/energy cover of the neutron process with permanent change of the time system.

The drive of the neutron collapse is the *Quantum Membrane*, loaded by the electron process. Electron process is work against the *Quantum Membrane*.

10A2
$$e_e = \frac{dmc^2}{dt_o\varepsilon_x}\sqrt{1-\frac{i^2}{c^2}}\left(1-\sqrt{1-\frac{(c-i)^2}{c^2}}\right)$$

Is it a real change? Yes, this is the load of the *Quantum Membrane* by the electron process: acceleration at constant speed, equal to $i = \lim a\Delta t = c$.

The *Quantum Membrane* is about the speed of quantum communication.

10A3
$$\frac{dmc^2}{dt_o}\sqrt{1-\frac{(c-i)^2}{c^2}}$$

The *inflexion point* establishes the values of the *quantum entropy*, function of the speed of quantum communication.

There are two distinguishing specifics of elements:
- the *intensity* of the electron process, characterising the relation between the proton and the neutron processes; and
- the *speed value of quantum communication*.

These two are connected and strongly relate to each other. The higher is the intensity of the electron process (the less is the value of the intensity coefficient within the electron process formula), the higher is the speed of quantum communication.

With regards to 10A3, the speed value of quantum communication is about the quantum entropy of the element as well.

Ref 10A3

With reference to 10A2, electron process *blue shift* conflict means: the developing *Quantum Membrane* of an element has more factors, than just the proton-electron-neutron processes.

Quantum Membranes are of different loads. The *Quantum Membrane* of elements with dominant neutron process intensity has straightforward acting character. There is no electron process *blue shift* surplus in elementary relations with neutron process dominancy. The proton process cover need of the loaded neutron process "swallows" all generating *blue shift* impacts in accordance with the proton/neutron intensity relation.

Elements with proton process dominancy (there are only just 8 within the periodic table: *Hydrogen, Helium, Oxygen, Nitrogen, Carbon, Sulphur, Calcium* and *Silicon*) have electron process *blue shift* surplus with a "challenging" *Quantum Membrane*, since their neutron process intensity is behind the proton process. Communication with these elements means *blue shift* conflict, since all of them have loaded *Quantum Membrane* with less intensive use.

10A4

There is no way for having *Quantum Membranes* of different intensity operating in parallel: *Quantum Membranes* consolidate their load.

With reference to 2C3 the value of the speed of quantum communication corresponds to the square of the relations of the quantum entropy gradients,

Ref 2C3

$$c_x = c_o \sqrt{\frac{qe_x}{qe_o}}$$

10A5

It is easy to accept that the resulting speed of quantum communication of different, but acting in parallel *Quantum Membranes* corresponds to an integrated value, established by the *blue shift* conflicts of the communicating systems. The higher is getting less, the lower is getting more. Quantities obviously impact the result in line with the proportions of the components.

Higher frequency of the *Quantum Membrane* means higher speed of quantum communication. There is no change in the function of elementary characteristics.

$$\varepsilon_x = \frac{\varepsilon_{px}}{\varepsilon_{nx}} \sqrt{1 - \frac{(c_x - i_x)^2}{c_x^2}} = \frac{dt_n}{dt_p}$$

10B1

The intensity coefficient of the electron process of the element is without change, remains the same just the appearance of these characteristics belongs to different – increased or decreased – speed values of quantum communication.

Increased speed of quantum communication increases the intensity difference between the proton and neutron processes, decreased speed makes the opposite, makes the difference less.

Increased quantum communication strengthens the dominancy of the processes. Less intensive weakens it.

10B2
$$\frac{dmc^2}{dt_p}\left(1-\sqrt{1-\frac{v^2}{c^2}}\right) >> or << \frac{dmc^2}{dt_o}\sqrt{1-\frac{(c-i)^2}{c^2}}\left(1-\sqrt{1-\frac{v^2}{c^2}}\right)$$

The proton process dominant *Oxygen* becomes more proton process dominant, with increased absolute surplus of the electron process;

Uranium with significant neutron process dominancy will have more weight as indicator of the increased intensity surplus of the neutron process.

10B3
$$\frac{dmc^2}{dt_o}\sqrt{1-\frac{v^2}{c^2}}\left(1-\sqrt{1-\frac{v^2}{c^2}}\right) = \frac{dm}{dt_o}\frac{\sqrt{c^2-v^2}}{1}\frac{c-\sqrt{c^2-v^2}}{1} = \frac{dm}{dt_o}\left(c\sqrt{c^2-v^2}-c^2-v^2\right)=$$

$$= \frac{dm}{dt_o}c^2\left[\frac{\sqrt{c^2-v^2}}{c}-1\right]+\frac{dm}{dt_o}v^2;$$ with growing c proton process dominancy is *increasing*

$$\frac{dmc^2}{dt_o}\sqrt{1-\frac{(c-i)^2}{c^2}}\left(1-\sqrt{1-\frac{v^2}{c^2}}\right) = \frac{dm}{dt_o}\frac{\sqrt{2ci-i^2}}{1}\frac{c-\sqrt{c^2-v^2}}{1} =$$

$$= \frac{dm}{dt_o}c\sqrt{2ci-i^2}-\sqrt{2ci-i^2}\sqrt{c^2-v^2} = A\frac{dm}{dt_o}c-A\frac{dm}{dt_o}\sqrt{c^2-v^2};$$ where $\sqrt{c^2} > \sqrt{c^2-v^2}$

with growing c neutron process dominancy is *increasing*!

The conclusion is that with the *growth* of the speed of quantum communication the dominances of processes grow, while with the decrease they are getting less.

With the increase of c the dominancy is strengthening!

As the speed of quantum communication is the same for the proton and neutron processes of an element, this change does not modify the relation of the two processes, the characteristic of the element, established by the electron process intensity of the element.

S.
10.1

10.1
Impact of *gravitation*

Ref.
9F4
9F5

The measured speed value of *Earth's* quantum communication is around 300,000 km/sec. With reference to 9F4 and 9F5 this is the reason of the gaseous and solid statuses of elements on the surface of the *Earth*, all with electron process surplus and proton dominancy.

Ref.
Table
9.1

With reference to Table 9.1, the status of *Hydrogen*, *Helium*, *Oxygen* and *Nitrogen* is gaseous as the increasing speed of their quantum communication, result of *Earth's gravitation*, strengthens their proton dominancy.

The other 4 elements (*Carbon, Calcium, Sulphur* and *Silicon*) also with proton dominancy are of solid status, as the higher speed value of their quantum communication is decreased to the speed of the quantum communication of *Earth's gravitation*. (And so, the dominancy becomes weakened.)

For all other elements of the periodic table with neutron process dominancy, the natural speed value of their quantum communication is higher than the speed of *Earth's gravitation*. Less speed value weakens the process dominancy and the energy/mass status of the element.

The best proof of the increased energy content of the above told (*Carbon, Silicon* but especially *Calcium* and *Sulphur*) elements is the heat generation when mixing them with water.

> Water represents the quantum communication of *Earth's gravitation*, as *Oxygen* and *Hydrogen* have been operating within this quantum system. Water, with its *Hydrogen* content of infinite neutron process demand and *blue shift* surplus communicates with these elements.
>
> In the case of mixing, there could not be various values of quantum communication in charge. The establishing integrated common quantum communication system will be corresponding to the energy content and the proportions of the mix – the frequency of the resulting *blue shift* conflict of the *Quantum Membrane*. The solution of these elements by water, the liquid status and heat generation is the proof they are integrating into the dictated by water common (and of less intensity) *Quantum System of Reference*.
>
> The new system is a *Quantum Membrane* with less speed value of quantum communication. The intensive heat generation during the mixing is the proof of the conflict between the quantum communication values and the *blue shift* impacts of the components.
>
> The developing heat is not just about the obvious *blue shift* conflict between elements, all with electron process *blue shift* surplus, but also about the fact that the easy soluble *Calcium* and *Sulphur* elements loose on their speed value of quantum communication.

Earth's gravitation, an electron process with sphere symmetrical expanding acceleration at constant speed for infinite time is result of the internal *blue shift* – mass/energy – energy/mass conflict of all elements of the *Earth*.

This conflict is the driving force of *Earth's* expansion: *gravitation*.

The acting *Quantum Membrane* at speed $i = \lim a\Delta t = c$ develops obvious resistance against the motion. The conflict goes back deep into the electron process (*Earth*) structure, components in expansion with energy surplus. The deep conflict is the *plasma* in the depths of the *Earth* itself, a non-identified conflicting stage: electron process *blue shift* conflict of infinite intensity with no definition of the conflicting elementary structures.

Ref.
S.8.3

As the *blue shift* impact of the sphere symmetrical expanding acceleration is working against the *Quantum Membrane*, components close to the conflicting surface become identified by their natural characteristics: Elements with proton, neutron and electron processes of various intensities!

Natural minerals, developing as result of the conflict are all in balanced energy/mass relation with each other, result of the natural establishment, keeping their quantum communication characteristics.

Ref. Table 9.1

Why do minerals keep elements in balanced state with different speed of quantum communication: for example *Uranium* at speed *378,163 km/sec*, *Gold* at *366,740 km/sec* and *Zirconium* at *339,451 km/sec*?
The reason is simple: this is the balanced status, the driving *blue shift* conflict of the *plasma*, the expanding and cooling status of *Earth gravitation* created at that depth and at that time point. And these statuses have been identified as different elements.

Gravitation, the sphere symmetrical expanding acceleration of *Earth* is product of the expanding potential of *Earth's* mass/energy internals.
In other words, the increased quantum speed of natural minerals is part of the driving force of *Earth gravitation*.
The *blue shift* conflict of the expanding *plasma* status is the engine, but all hardened minerals deep in *Earth* are part of it. Reserves are deep inside.
Gravitation is *blue shift* impact with energy loss.
Neutron dominancy is result of the loosing *blue shift* "capacity" of the expanding acceleration (of *Earth's gravitation*). Natural minerals, with neutron process dominancy become local unique energy/mass reserves at the surface of the *Earth*.

The speed value of quantum communication of minerals remains the same without change – while they have been kept in natural conditions – with the only difference that the solid structure experienced within the core of the *Earth* during mining might be turning into powder.
The reason is the difference in the speed value of the quantum communication of the mineral and *gravitation*.

> The *Quantum System of Reference* within the structure of the minerals is with neutron process dominance and the *Quantum Membrane* of the mineral is with higher speed of quantum communication than that on the *Earth* surface. The external boundaries of this *Quantum Membrane* communicate with a system of less intensity. As the speed of quantum communication of the *Quantum Membrane* outside the minerals is of less intensity, with less frequency, with less support from, the solid structure remains solid, but powder like. At the same time the element of the mineral has no *blue shift* surplus.

Natural minerals are balanced mixture of elements. The balance has been established between elements as result of natural quantum communication.

Increased impact of *Earth's gravitation* changes the speed of quantum communication of the elementary components of the mineral without destruction of the natural internal balance of minerals.

The speed value of quantum communication on the *Earth* surface is less than the natural quantum speed of the elements of the mineral deep in the *Earth*. The intensities of the impact of *gravitation* are different.

As first indicator of the change is the measured reduced weight of the element, which is understandable as neutron collapse becomes of less intensity.

Increased intensity means in practice the melting of the minerals.
Melting is increased *blue shift* conflict within the mineral, generated by external *blue shift* impact source.

external *blue shift*   internal *blue shift*, acting in   the resulting
impacting   the *Uranium* element   conflict

$$X\frac{dmc^2}{dt_i\varepsilon_x}\left(1-\sqrt{1-\frac{(c-i)^2}{c^2}}\right) + U\frac{dmc_U^2}{dt_i\varepsilon_U}\left(1-\sqrt{1-\frac{(c_U-i_U)^2}{c_U^2}}\right) = \qquad \Big\downarrow$$

$$= X\frac{dmc_c^2}{dt_i\varepsilon_c}\left(1-\sqrt{1-\frac{(c_c-i_c)^2}{c_c^2}}\right)$$

10B4

c in the formula means *conflicting*

The transformation of the quantum speed happens during melting.

c_c- the conflicting common speed value is close to the speed of quantum communication of *gravitation* on the surface of the *Earth* in line with the proportions of the *blue shift* impact of the conflict. (The *blue shift* impact of the "heating" source is much more dominant in the conflict.)

> ➤ Liquid (melted) state is result of increased *blue shift* conflict (as result of external impact – fire);
> ➤ Solid status again is result of cooling (taking off the electron *blue shift* surplus), result of the quantum communication with the *Quantum Membrane* of *Earth's gravitation*.

Melted and hardened minerals and elements become part of the quantum system and membrane of *Earth's gravitation*.

Gravitation takes off *blue shift* and the element becomes solid again, but with serious consequence: the speed of its quantum communication corresponds to less value, equal to the speed value of *Earth's gravitation*.
This is the reason why elementary components behave differently as part of minerals and products of technology preparations: the reason is the change (reduction) in the speed value of their quantum communication.

S.   **10.2**
10.2   **Cooling by *Earth gravitation***

All elements and minerals are special gifts of the nature. Heavy metals are with energy reserve. This is not just about the *Uranium* or *Plutonium* as fuel of nuclear reactors but all other elements in the second half of the *Periodic Table.*

Ref.
S.4   With reference to Section 4, fission of heavy elementary structures results in elements with higher proportion of the proton process – with the potential benefit of the generation and use of the developing electron *blue shift* surplus.

But nuclear fission destructs the elementary structure.

Heavy elements work with increased speed of quantum communication. In the case of communication with *Earth's gravitation* (cooling after melting), heavy elements certainly loos on their speed of quantum communication. The benefit of the communication is the potential use of the difference – without damage of the elementary structures.

While higher speed of quantum communication means higher speed of $i = \lim a\Delta t = c$ *blue shift* impact, this fact does not result in longer time frame of the electron the process – in the contrary, it results in higher intensity!

As it is also proven by the increased intensity of the inflexion point of these elements. How can this happen?

The explanation also gives answer to the question, why electron processes communicate?

Electron process:

10C1
$$\frac{dmc^2}{dt_i \varepsilon_x}\left(1 - \sqrt{1 - \frac{(c-i)^2}{c^2}}\right) = \frac{dmc^2}{dt_o \varepsilon_x}\sqrt{1 - \frac{i^2}{c^2}}\left(1 - \sqrt{1 - \frac{(c-i)^2}{c^2}}\right);$$

10C2 i as being $i = \lim c$ it gives quasi equal values in any relations of $\lim(c_x - i_x) = 0$ and $\lim(i_x/c_x) = 1$:

10C3
$$\sqrt{1 - \frac{i^2}{c^2}} \cong \sqrt{1 - \frac{i_1^2}{c_1^2}} \cong \sqrt{1 - \frac{i_2^2}{c_2^2}} \cong \dots \cong \sqrt{1 - \frac{i_{n-1}^2}{c_{n-1}^2}} \cong \sqrt{1 - \frac{i_n^2}{c_n^2}}$$

This is the reason *blue shift* impacts of different elements may have surplus, conflict and deficit, elements communicate with each other.

At the same time the difference in the speed values of their quantum communication might be significant.

10C4   This is the reason the intensity of the quantum entropy gradient, acting at the inflexion point de facto depends only on the speed value of quantum communication.
$$qe = \frac{dmc^2}{dt_o}\sqrt{1 - \frac{(c-i)^2}{c^2}}$$

$$dt_{ix} = \frac{dt_o}{\sqrt{1 - \frac{i_x^2}{c_x^2}}} = const \qquad i_x = \lim a_x \Delta t = c_x \quad \text{and} \quad \lim \frac{i_x}{c_x} = 0 ;$$

result in constant value for any i_x variant.

10C5

And so, the time system for all electron processes is one and the same, while the intensity of the *blue shift* impact is changing as the speed of quantum communication varies.

This gives clear explanation to the quantum communication of elements and is in line with the experience and practice.

$$\lim(c_x - i_x) = 0 ; \quad \text{but } c_x \text{ is variant,}$$

$$e_x = \frac{dmc_x^2}{dt_{ix}\varepsilon_x}\left(1 - \sqrt{1 - \frac{(c_x - i_x)^2}{c_x^2}}\right) \qquad \text{value in } \sqrt{1 - \frac{(c_x - i_x)^2}{c_x^2}} \text{ also varies,}$$

10C6

with the growth of c_x is getting less

e_x depends on the value of c_x in the nominator; with its growth e_x increases.

What are the consequences?
- neutron processes of all elements and all natural relations are driven at constant time system, but with *various* speed values of quantum communication;
- a *Quantum System of Reference* cannot exist/operate/communicate with different quantum speed values – speed values are consolidating;
- quantum communication of elements means establishment of common (integrated) quantum speed;
- ε_x of the element is without change, but proton and neutron processes happen at common integrated speed of quantum communication.

Earth's gravitation as *blue shift* impact and surplus is open for elementary communication and communicates with elementary processes.

Heavy metals are with increased speed of quantum communication.
There would be a benefit if this speed surplus could be used and utilised.
How can a speed surplus be used?
In the case of melted (liquid) state of heavy metals, the cooling is the way of communication. In this case the *blue shift* surplus of the liquid status is taken off. The cooling process goes with the speed of quantum communication of *Earth's gravitation*. The melted status is a *blue shift* conflict of increased frequency. Now this frequency comes down as the cooling dictates.
The cooling is the benefit, because the cooling is deeper (with less speed of quantum communication) than the status of the metal was when the heating actually started.
Heating starts at the speed value of the quantum communication of the heavy metal, the cooling however ends at the speed value of the *gravitation* of *Earth*.

$$\frac{dmc_{hm}^2}{dt_o\varepsilon_{hm}}\sqrt{1-\frac{i_{hm}^2}{c_{hm}^2}}\left(1-\sqrt{1-\frac{(c_{hm}-i_{hm})^2}{c_{hm}^2}}\right)+$$

10D1
$$+X\frac{dmc_E^2}{dt_o\varepsilon_E}\sqrt{1-\frac{i_E^2}{c_E^2}}\left(1-\sqrt{1-\frac{(c_E-i_E)^2}{c_E^2}}\right)=$$

10D2
$$=Y\frac{dmc_{com}^2}{dt_o\varepsilon_{hm}}\sqrt{1-\frac{i_{com}^2}{c_{com}^2}}\left(1-\sqrt{1-\frac{(c_{com}-i_{com})^2}{c_{com}^2}}\right)$$

Heating, with reference to 10D1 results in melted state, with reference to 10D2 in certain common speed of quantum communication. The quantum speed of this melted state, as expected is already less than the natural speed of the quantum communication of the heavy metal in its mineral status.
The liquid state of the metal depends on the *blue shift* surplus of the conflict, which function of the external *blue shift* impact of heating.

It is important to note that for making the heavy metal melted the speed of quantum communication of the heating source does not need to be higher than that of the metal's. Creation of the *blue shift* conflict is the key.
The conflict creates its quantum communication speed within the melted heavy metal structure.

Cooling goes with quantum communication of *Earth's gravitation*:

$$Y\frac{dmc_{com}^2}{dt_o\varepsilon_{hm}}\sqrt{1-\frac{i_{com}^2}{c_{com}^2}}\left(1-\sqrt{1-\frac{(c_{com}-i_{com})^2}{c_{com}^2}}\right)-$$

10D3
$$-Z\frac{dmc_E^2}{dt_o\varepsilon_E}\sqrt{1-\frac{i_E^2}{c_E^2}}\left(1-\sqrt{1-\frac{(c_E-i_E)^2}{c_E^2}}\right)=$$

10D4
$$=\frac{dmc_E^2}{dt_o\varepsilon_{hm}}\sqrt{1-\frac{i_E^2}{c_E^2}}\left(1-\sqrt{1-\frac{(c_E-i_E)^2}{c_E^2}}\right)$$

The beauty of 10D3 is that *Earth's gravitation* represents an infinite portion within the communication. This way, while the speed value of the melted stage is getting less and less the cooling happens under the quantum impact of *gravitation* with constant quantum speed and constant frequency. (The decrease of the quantum speed of one of the components does not result in the increase of the other.)

As consequence, the speed value of the quantum communication of the liquid status is decreasing, until the heavy metal reaches its elementary solid structure with no *blue shift* surplus and so, with no need for cooling any more.
At this stage however the establishing by the cooling speed of quantum of communication of the heavy metal is equal to that of *Earth's gravitation*.

The final result and the *benefit* of the process is that the energy taken by cooling is more than the energy used for melting!

With reference to 10D1 and 10D3

Ref.

$$X \frac{dmc_E^2}{dt_o \varepsilon_E} \sqrt{1 - \frac{i_E^2}{c_E^2}} \left(1 - \sqrt{1 - \frac{(c_E - i_E)^2}{c_E^2}}\right) < Z \frac{dmc_E^2}{dt_o \varepsilon_E} \sqrt{1 - \frac{i_E^2}{c_E^2}} \left(1 - \sqrt{1 - \frac{(c_E - i_E)^2}{c_E^2}}\right) \qquad \begin{matrix}\text{10D1}\\\text{10D3}\end{matrix}$$

as

Ref.

$$\frac{dmc_{hm}^2}{dt_o \varepsilon_{hm}} \sqrt{1 - \frac{i_{hm}^2}{c_{hm}^2}} \left(1 - \sqrt{1 - \frac{(c_{hm} - i_{hm})^2}{c_{hm}^2}}\right) > \frac{dmc_E^2}{dt_o \varepsilon_{hm}} \sqrt{1 - \frac{i_E^2}{c_E^2}} \left(1 - \sqrt{1 - \frac{(c_E - i_E)^2}{c_E^2}}\right) \qquad \begin{matrix}\text{10D1}\\\text{10D3}\end{matrix}$$

The benefit is proportional to the quadrat of the quantum speed difference:

$$b = \frac{c_{hm}^2}{c_E^2} \qquad \text{10D5}$$

In the case of *Uranium* clean ore the energy benefit of cooling is *58%*!

Melting point of *Uranium* is $1132.2\,^{\circ}C$ or 1405.3 K.

S.
11

11
<u>Gravitation is quantum *treasure*</u>

The soil on *Earth* surface is a specific mass/energy composition of minerals. The status of soil has not been clearly defined, but it is with elementary processes, quantum communication and *Quantum Membrane*.

Ref.
S.9.5

With reference to Section 9.5, *Earth's plasma* is in *blue shift* conflict of infinite intensity. This *blue shift* conflict is the source of *Earth gravitation*, its sphere symmetrical expanding acceleration at constant speed

$$i = \lim g\Delta t = c$$

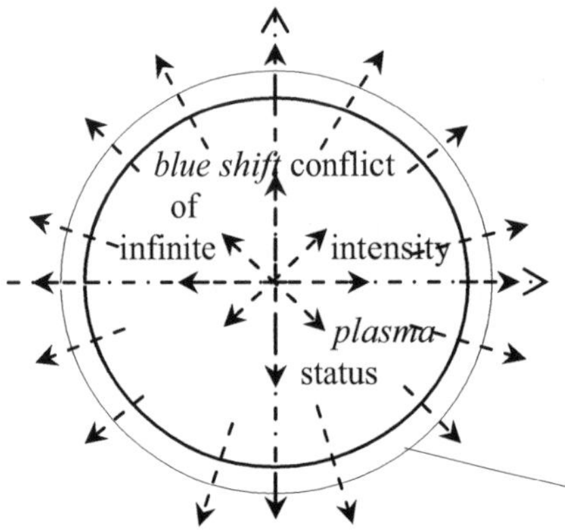

Plasma status:
Increased *blue shift* conflict; sped of quantum communication is of infinite value.

In conventional terms:
- inner core;
- outer core;
- mantle;
- crust (in 40-50 km depth)

Fig.
11.1

Fig.11.1

The *blue shift* conflict and the speed of quantum communication from the *plasma* towards the surface are of less and less value. The hardening of the *Earth* structure means less *blue shift* conflict and less value of quantum speed.
Gravitation means a kind of "cooling" process – utilisation of the intensity of the electron process *blue shift* impact coming from the plasma state. The closer the elementary components of soil minerals to the *Earth* surface are the more have been lost from the original intensity of the *blue shift* impact and the quantum speed value of the *plasma*. The resulting $c \cong 300,000$ km/sec quantum speed at the surface is the one *Earth* is communicating.

Ref.
11A1

The speed of quantum communication is the only parameter soil components can lose = resulting in *blue shift* loss.

$$e = \frac{dmc^2}{dt_i \varepsilon_x}\left(1 - \sqrt{1 - \frac{(c-i)^2}{c^2}}\right)$$

Elements have been created in line with the acting speed value of quantum communication. Each element within the *Earth* has its own elementary balance in accordance with the cooling effect of *gravitation*.

Ref.
Tab.
9.5

The closer the mineral to the surface is the less is the value (with reference to Table 9.1 of Section 9.5) of its quantum speed, but still above $c \cong 300,000$ km/sec, the speed of quantum communication of *gravitation*.

Gravitation is permanent and the loss of the *blue shift* impact of the *plasma* through the surface of *Earth* is also permanent. The elementary structure of minerals in *Earth* corresponds to the established in that region speed values of quantum communication and to the intensities of the *blue shift* impacts.
It means: minerals with elements within the *Earth* represent the established quantum communication between the *plasma* and the *Quantum System of Reference* around the *Earth*.

11.1
***Gravitation* is electron function**

S.
11.1

Earth's plasma blue shift conflict of infinite intensity is the drive of the sphere symmetrical expanding acceleration of the *Earth*. The speed value of the quantum communication of *Earth plasma* is: $\lim c_{plasma} = \infty$.
This keeps *Earth* in sphere symmetrical expanding acceleration.

The value of the quantum speed of the *blue shift* impact approaching from plasma state towards the surface is decreasing, as the conflict of the *plasma* through the expansion (*gravitation*) is resolving.
Elements of minerals deep in the *Earth* have increased intensity of electron process *blue shift* impact. This is well in line with the increased speed value of quantum communication.

Plasma with its *blue shift* conflict is representing classical electron function with sphere symmetrical expanding acceleration. The conflict within the *plasma* is acting as the drive of the expansion.
While the *blue shift* is acting, the components of the *plasma* and also of the *Earth's* crust are subjects to expansion and acceleration.

Electron process is transformation of mass/energy into *blue shift*. With the expansion process (*gravitation*) from plasma state towards the surface the conflict is less and less and the *blue shift* loss is more and more.

The speed of the quantum communication of the expansion process varies, establishing "elements" of different kinds.
In deeper layers the quantum speed is higher. The speed value is decreasing towards the surface. Elements within the *Earth* have their specific internal elementary processes with *Strong* and *Weak Interrelations*.

The intensity of the elementary electron process *blue shift* impact is:

11A2 $$e_e = \frac{dmc_x^2}{dt_i \varepsilon_x}\left(1 - \sqrt{1 - \frac{(c_x - i_x)^2}{c_x^2}}\right) = \frac{dmc_x^2}{dt_o \varepsilon_x}\sqrt{1 - \frac{i_x^2}{c_x^2}}\left(1 - \sqrt{1 - \frac{(c_x - i_x)^2}{c_x^2}}\right)$$

x indexes relate to the certain element.

Earth crust is full of these elementary electron processes. The sphere symmetrical expanding acceleration (*gravitation*) of *Earth* establishes all elementary processes.

The *Periodic Table* shows the characteristics of all elements. Elements are parts of *Earth* minerals. Their quantum speed of communication very.

> In the case of *Oxygen, Hydrogen, Nitrogen, Helium* and *Carbon* (as specific element), the speed corresponds to $c \cong 300,000$ km/sec.

> In the case of all other elements the speed is different. If impacted however on the surface of the *Earth*: boiling, steaming or condensing, the consolidated speed will be changing to $c \cong 300,000$ km/sec.

The permanent established loss on *blue shift* impact means *Earth* surface is with electron process *blue shift* deficit. Electron process *blue shift* impacts are welcome and have been swallowed by *Earth* in infinite quantities.

S.
11.2

11.2
The benefit of *gravitation*: quantum speed difference

Earth is in permanent loss of its internal *blue shift* impact because of *gravitation*. The established intensity balance of proton and neutron processes of elements in *Earth'* natural minerals correspond to the impact of *gravitation*.

The deeper is the element to be found in *Earth'* crust, core, soil,

(1) the higher are the intensities of the electron and neutron processes;
(2) the higher is the speed value of quantum communication;
(3) the wider is the gap between the intensities of the proton and neutron processes of the elementary balance; and
(4) the higher is the dominancy of the neutron process.

Blue shift drive assistance (of low intensity) is coming from all those elements surrounding us with *blue shift* surplus: *Hydrogen, Oxygen, Nitrogen, Helium, Carbon, Calcium, Sulphur* and *Silicon.*

Soil is of full energy potential, but without the necessary initiation (*blue shift* surplus) of the process. *Oxygen, Hydrogen, Nitrogen, Carbon,* water and a couple of other elements are the drives and we, ourselves and the vegetation around us are the ones establishing the life-contact

Blue shift conflict from *plasma* is weakening, *Earth's* crust is hardening.

If minerals are melted, the conflict is increased again, just at a lower speed of quantum communication. Elements in liquid elementary structure flow out. During the process of hardening – losing on *blue shift* conflict again – the establishing elementary structure is hardening at the speed value of the quantum communication on the surface of *Earth*.

There are parts of minerals, which cannot be melted on the surface of the *Earth*. The reason is that the intensity of the *blue shift* conflict and the value of the quantum speed are insufficient for melting these minerals.

The intensity of the *Quantum Membrane* depends on the actual speed of quantum communication of the quantum entropy gradient.

$$\frac{dmc_x^2}{dt_o}\sqrt{1-\frac{(c_x-i_x)^2}{c_x^2}}$$

11B1

Earth's plasma is of $\lim\dfrac{dmc_{pl}^2}{dt_o}=\propto$ energy intensity.

11B2

The intensities in 11B1 and 11B2 could be written relating to dt_i as well, the meaning would not be changing just the numerical descriptions of the same physical speed values would be different.

The quantum communication establishing the balance between *Earth's plasma* and the *Quantum Membrane* around *Earth* surface is permanent.

Quantum communication means, *Earth's plasma* is providing electron process *blue shift* impact (*gravitation*) to the external *Quantum System of Reference* of the *Earth*: keeps the external *Quantum Membrane* loaded.

Earth' plasma is electron process *blue shift* conflict of infinite intensity. The external *Quantum Membrane* is of infinite intensity need:

$$e_E=\frac{dmc_{pl}^2}{dt_o}-\frac{dmc_x^2}{dt_o}\sqrt{1-\frac{(c_x-i_x)^2}{c_x^2}}=\frac{dm}{dt_o}\left[c_{pl}^2-\sqrt{1-\frac{(c_x-i_x)^2}{c_x^2}}\right]$$

11B3

(1) the need in quantum impact of the *Quantum System of Reference* surrounding *Earth* is of infinite value;
(2) the speed of quantum communication within the *plasma* is also of infinite value;

The equation in 11B3 describes an infinite process with gradual decrease of the speed of the *blue shift* impact from c_{pl} within the *plasma* state to c_x at the surface of the *Earth* – for reaching equality and balance.

This is the reason of *Earth's* permanent electron process *blue shift* impact – *gravitation* – with decreasing acceleration value (*g*) and increasing time (Δt) components: $i=\lim g\Delta t=c$.

$$w=dm\frac{(c_{pl}^2-c_x^2)}{dt_o}=\propto$$

The work of the intensity difference, while the impact is reaching the surface is distributing within the *Earth's* core.

11B4

This is the reason of having minerals with elementary compositions of different speed values of quantum communication towards the *plasma*.

The speed difference between the *Earth's* surface and crust points towards *plasma* is developing, elements and minerals are to be found accordingly.

Ref.
Table
9.1

While the speed of quantum communication on the surface *Earth's* is value of $c_E \cong 300,000$ km/sec, with reference to Table 9.1 at the depth of *Uranium* mining, this speed is $c_U = 378,000$ km/sec, at the depths of *Gold* and *Cuprum* are $c_G = 366,000$ km/sec and $c_{Cu} = 327,000$ km/sec.

S.
11.3

11.3
Acceleration at constant speed

Plasma within the *Earth* is of infinite *blue shift* conflict: a *Quantum Membrane* of infinite intensity. The drive of the sphere symmetrical expanding acceleration of *Earth* is coming from the *Quantum Membrane* of the *plasma*, an electron process *blue shift* conflict of infinite intensity.

Earth is expanding and this way loading the *Quantum Membrane* by its developing *blue shift* impact, *gravitation*. This *blue shift* impact, keeping *Earth's Quantum Membrane* loaded is a loss. The infinite *blue shift* conflict of the *plasma* compensates the loss caused by the expansion (*gravitation*).
The effect of *gravitation* is balanced by the energy of the *plasma*.
The consolidated speed of the sphere symmetrical expanding acceleration of the *Earth* is constant and equal to $i = \lim g\Delta t = c$.

Earth is in electron function: sphere symmetrical expanding acceleration at quasi constant speed.

The intensities of neutron processes within the *Earth* core, crust vary in line with the specific elementary balance, established by the intensity of the *blue shift* impact of *gravitation*. The electron process intensity coefficient and the relation of proton and neutron process intensities follow the change.

Proton processes continue in electron processes; the intensity of the acting electron process *blue shift* impact at certain *Earth* depth level corresponds to the acting at that level intensity and quantum speed value of *gravitation*. Electron processes turn into neutron processes, in line with the intensity of the *blue shift* impact of *gravitation*. This way neutron processes deep in *Earth* might have massive electron process *blue shift* drives.

Elements have been born this way: with the variety of the proton/neutron process relations, corresponding element by element to the actual acting intensity of the *blue shift* impact at that certain depth level of *Earth*' crust.
Elements with massive or full *blue shift* loss are on the surface and close to the surface.

Earth is losing on its *blue shift* impact.

As the intensity of the *plasma* is decreasing, the status of the core and the surface at first is liquid, after solid.
The *plasma* is shrinking towards the centre. *Blue shift* impacts of the conflict as *gravitation* propagates through the *Earth* crust are establishing elementary compositions of decreasing electron process intensity.

Earth is accelerating while the position of the surface is constant and the speed value of $i = \lim g\Delta t = c$ is also constant.
The acceleration (*gravitation*) is decreasing, the time flow is lengthening.

The *blue shift* impact of *plasma* is "producing" elementary compositions within the crust with variety of quantum speed values and intensities of *blue shift* impact: the closer the composition to the *plasma* is, the higher is the speed of the operating quantum communication.

The core and the crust are connecting the *Quantum Membrane* of the *plasma* with the "external" *Quantum Membrane* of the *Earth'* surface.
The *blue shift* impact at *plasma* level is initiated as of infinite high quantum speed and of infinite high electron process intensity.

$$e_{plasma} = \frac{dmc_\infty^2}{dt_i}\left(1 - \sqrt{1 - \frac{(c_\infty - i_\infty)^2}{c_\infty^2}}\right)$$ 11C1

The intensity of the external *blue shift* impact is:

$$e_x = \frac{dmc_x^2}{dt_i}\left(1 - \sqrt{1 - \frac{(c_x - i_x)^2}{c_x^2}}\right)$$ 11C2

The speed values of the quantum communication of elements within the *Earth* are connecting the two speed values within 11C1 and 11C2.
Minerals with elements are formulating at certain intensity values of quantum impact and speed of quantum communication.
To be noted:

   whatever are the speed values of the quantum communication, with reference to 10C3 the time system of the *blue shift* impact is always the same;

   ε_x the coefficient of the electron process has not been shown in equations 11C1 and 11C2, as the analysis is of global level. It is necessary to take it into account if there are elements to be specified at different levels and depths.

In line with 11C2 and 10C3, electron processes are communicating, and the work intensity value between *plasma* and the external *Quantum Membrane* in any internal point of the *Earth* is:

$$\Delta e = e_{plasma} - e_x = w$$ 11C3

11C4   Taking for $\lim\left(1-\sqrt{1-\dfrac{(c_\infty-i_\infty)^2}{c_\infty^2}}\right) \cong \lim\left(1-\sqrt{1-\dfrac{(c_x-i_x)^2}{c_x^2}}\right)=1$

The energy intensity difference and the work intensity are proportional to the difference of the square of speed values of quantum communication.

11C5

$$w_x = \frac{dm}{dt_i}\left(c_\infty^2 - c_x^2\right)=\infty$$

11C5 means permanent propagation of *blue shift* impact from *Earth's plasma* of infinite intensity through different depth levels of the *Earth*.
The speed of quantum communication is decreasing towards *Earth's* surface. Its value depends on the depths of the element within *Earth's* crust.

The speed of quantum communication of the *Quantum Membrane* on the surface of the *Earth* is $c \cong 300{,}000$ km/sec;

the acceleration is $a = g = 9.81\ m/\sec^2$, decreasing;

the intensity of the *blue shift* impact from *Earth's plasma* is of infinite value.

Elementary processes in the crust are being created in line with the value of the quantum speed and the intensity of the *blue shift* impact.

The extremely high intensity of the quantum impact from *plasma* increases the intensity of elementary processes around the *plasma*. But elementary processes are subjects to "cooling" by *gravitation* from the side of the external *Quantum Membrane*. Therefore intensity values are decreasing.
So,

11D1   the intensity difference is: $\Delta e_{plasma} = e_{plasma} - e_x$;

11D2   the intensity release by gravitation is: $\Delta e_g = e_x - e_g$;

11D3   the full difference between *plasma* and *Earth's* surface is: $\Delta e = e_{plasma} - e_g$

The acting conflict and the impact of the *plasma/gravitation* energy transfer process can be assessed/controlled by the temperature at different depths and crust levels of the *Earth*.

S.
11.4

11.4
Quantum treasure = gravitation

Soil on the surface of *Earth* is filled with elementary proton and neutron process components, but without the sufficient corresponding electron process *blue shift* drive.
Because of the missing electron process *blue shift* drive elementary processes cannot be completed on the *Earth's* surface, leaving "soil" behind.
Soil is representing a composition, where neutron processes remain without sufficient electron process intensity support and *blue shift* drive.

Uncompleted elementary processes with electron process *blue shift* deficit within the soil, as consequence of *gravitation* mean proton process dominance at the surface.

Why proton process dominance? Because
- the quantum speed on the way from *plasma* to the surface is decreasing;
- the intensity of the electron process *blue shift* impact on the way from *plasma* to the surface is established in harmony with the impact of *gravitation* and
- getting less and less towards the surface;
- generation of electron processes means completed proton processes;
- neutron process is neutral as it always has to be driven;
- neutron processes are getting less and less driven.

The general proton *process* dominance of the *Earth's* surface, soil, crust, core does not contradict to the *intensity* dominance of the neutron processes in different depth levels. The intensity dominance of the neutron processes in the core means increased intensity because of the increased quantum speed.

One of the most important findings is the negative electron process *blue shift* gradient acting in *Earth* core, crust and soil. *Earth* with permanent loss of *blue shift* impact through *gravitation* is in electron process *blue shift* need – as everyday practice clearly proves it!

$$e_g = \frac{dmc^2}{dt_i}\left(1 - \sqrt{1 - \frac{(c-i)^2}{c^2}}\right) \qquad \text{11E1}$$

Gravitation provides *blue shift* impact to all mass/energy processes in *free fall* towards the *Earth* surface:

dt_i is the time system; dm is with general meaning;
c is the speed of quantum communication, measured on *Earth* surface.

In line with conventional physics, *weight* is $F = mg$; and $\qquad$ 11E2
energy potential (work) at h, relative to the *Earth* surface is: $W = mgh$ $\qquad$ 11E3

The *energy intensity potential,*
without motion in at level h, gives: $\qquad e_h = \frac{dm}{dt_i}gh \qquad$ 11E4

resulting in:
$$\frac{dm}{dt_i}gh = \frac{dmc^2}{dt_i}\left(1 - \sqrt{1 - \frac{(g\Delta t)^2}{c^2}}\right) \qquad \text{11E5}$$
where Δt is coming from $h = (g/2)t^2$

The energy/mass intensity potential of an "object" at h meters height above the surface in *free fall* with mass flow of dm / dt_i, is equal to 11E5.

If the "object" is down to the *Earth* surface in one single piece, the work intensity value in 11E5 has been utilised in its 11E3 mechanical format:

$$mgh = mc^2\left(1 - \sqrt{1 - \frac{(g\Delta t)^2}{c^2}}\right) \qquad \text{11E6}$$

Objects in free fall remain without the drive of the sphere symmetrical expanding acceleration of *Earth gravitation*. Objects are slowing down by $g = 9.81...m/\sec^2$ negative acceleration relative to the $i = \lim g\Delta t = c$ speed of *gravitation*. The slowdown process means they do not have the accelerating energy of *Earth gravitation* anymore. The lost work values of the "mechanical" and the *blue shift* impact correspond to 11E5.

Ref.
11E5

The mechanical effect can be measured at the moment of the landing.

There is a quantum effect in force, impacting objects in free fall different than the lost *blue shift* impact of *gravitation*. This is consequence of the interruption of *Earth* magnetic lines between the *North* and the *South* poles by the objects in free fall. The magnetic field is constant while *Earth* is in *gravitation* and rotation. Its stability is guaranteed by the balance of the acceleration and the *blue shift* loss of *gravitation*.

Ref
S.8

Objects, not having the accelerating *blue shift* impact are crossing magnetic lines in their free fall. Crossing magnetic lines, with reference to Section 8 results in electron process *blue shift* generation within the objects.

This quantum effect however is so minimal, that it is very difficult to measure it.

For having measureable quantum impact, the mass of the object in free fall shall be pulverized into infinite numbers of mass pieces. This ensures that the *Strong Interrelation* of *Earth* magnetic lines impact the internal electron process *blue shift* drive of the elementary processes of each mass piece.

One piece object in free fall crosses magnetic lines the same way but
(1) the integrated internal elementary relations and the overall balance is more stable; and (2) the pulverised mass has increased cross surface.
The interruption of magnetic lines causes *blue shift* conflict and surplus within the internal elementary process of the object, equal to generation of *electricity*.

Gravitation is also impacting *blue shift* signals released at certain height level, approaching *Earth* surface:
(1) the intensity of the *blue shift* impact is increased as being in conflict with *gravitation* within the *Quantum Membrane* above *Earth* surface;
(2) the *blue shift* impact of increased intensity is "mechanically" *blue shifted* by the collision with the *Earth* surface, speed of $i = \lim g\Delta t = c$;
   (The surface is proton dominant, taking electron process *blues shift* impacts, but only by the elementary processes of the internal structure.)
(3) the *blue shifted* by the *Earth's* surface impact is part of the *Quantum Membrane* of the conflict above the surface of increased intensity;
(4) the *blue shifted blue shift* is approaching its release point with increased intensity either to be *blue shifted* and turned back again or taken as *red shift* impact.

11.4.1. *If free fall of pulverized mass format is considered as benefit*

- In the case of elements with high quantum speed value (like *Uranium*, $c_U = 378,000$ km/sec) the conflict caused by the impact of *Earth* magnetic lines is developing slowly. The reason is the significant neutron process dominance of the element. It takes time and number of cycles reaching electron process *blue shift* conflict and surplus. If however is established, it is of high intensity.

- In the case of elements close to balance state (as clean *Carbon*, $c_C = c_{Earth} \cong 300,000$ km/sec) the conflict and the surplus is immediate, but of less intensity, as the speed of quantum communication is of less value.

> The relation of the intensities of the acting *blue shift* impacts of *Uranium* and *Carbon* powders is proportional to the square of their quantum speed values and is equal to *1.59* in the favour of the *Uranium*.
>
> The same relation of *Uranium* to *Cuprum* element is: *1.33*.
>
> Above relations mean that the intensity of the quantum impact within the *Uranium* element is 59% more than within clean *Carbon* and 33% more than that within *Cuprum*.

In conventional electricity generation practice *blue shift* impact is generated within the *Cuprum* cables of rotors impacted by the *Strong Interrelation* of magnetic poles. The impact is transferred by the speed of the quantum communication of the cable.

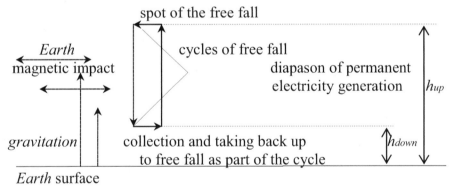

Fig.11.2
Fig. 11.2

With reference to Figure 11.2, pulverized mass object, in free fall, crossing *Earth* magnetic lines can generate electricity. The parameters of the generating electron process *blue shift* flow depend on the energy capacity of *gravitation* (the length of the free fall), the intensity of the magnetic effect, the quality and the quantity of the elements of the mass powder.

The impact of *gravitation* is twofold:
- the quantum impact is transforming into electricity;
- the mechanical impact gives work value, corresponding to 11E6.

S.
11.4.
2

Ref.

Fig.

11.3

11.4.2. If blue shift energy format is considered

The developing *Quantum Membrane* of limited size but of increased intensity is one of the benefits; the other is the *red shift* (electron process impact) – generating electricity at the reception point.

Quantum Membrane of increased intensity is result of the conflict of the *blue shift* impact and *gravitation*.

Nature provides variety of options for the cover of the *Earth* surface.
A soil, rich in *Silicon* is one of the best for reflection and *blue shifting blue shift* impacts. It is behaving different way than other soil components.
The *Quantum Membrane* of increased intensity, limited size and full of *blue shift* conflict with the *Earth's* surface might have its mechanical lifting effect to be used as well.

red shift impact is taken as *electricity flow* (1)

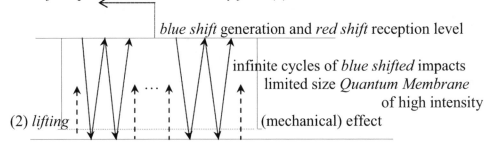

blue shift generation and *red shift* reception level

infinite cycles of *blue shifted* impacts
limited size *Quantum Membrane*
of high intensity
(2) *lifting*   (mechanical) effect

11F1   *Earth* surface with *blue shift* impact, speed of $i_{Earth} = \lim g\Delta t = c$

11F2   time system is *infinite long*, as: $dt_{Earth} = \dfrac{dt_o}{\sqrt{1 - \dfrac{i^2_{Earth}}{c^2_{Earth}}}}$

Fig.
11.3   Fig.11.3

The examples with the *Uranium* and *Carbon* elements and with the *blue shifted blue shift* impact above have not been experienced in industrial circumstances.
The goal is to demonstrate: *gravitation* is an energy source of infinite capacity with options for us to investigate, develop and use.

Ref.

A2

With reference to Attachment A3, the energy of *rain* has been measured. The generated electricity flow, value of up to 1.9 μA is result of *Earth gravitation* and the impact of *Earth* magnetic lines.

Nature provides infinite varieties of options to utilise the gift of
gravitation: our *quantum treasure*!

ATTACHMENTS A

Finding *Higgs boson*
/assessment on process based approach/

Higgs boson is not a particle rather a *quantum impact.*
Higgs boson is the integrated effect of the *Strong Interrelation.*
The measured in LHC experiments *Higgs boson* is the quantum impact of the disrupted and newly established *Strong Interrelation* of particles (processes) of the collision.

Background

The assessment is prepared on the basics of the mass-energy balance and the entropy principle of elementary processes. The key assumption is that elementary particles are *all* processes.
Definition of *time* can only be given if particles are taken as processes. Results of high speed collisions cannot be explained and described in correct way without taking into consideration the time impact.
Relativity is fully applied.
Protons, neutrons and the electron function with *electrons, muons* and *tau* are processes; as are also *quarks* of all flavors as well.
Strong Interrelation means the mass-energy balance of the proton and neutron processes under the control of the electron function.
Electron function (including *electron, muon* and *tau* processes) represents the *Weak Interrelation*, impacting the *Quantum System of Reference*, loading it into *Quantum Membrane.*

(Energy/mass) quantum is entropy generated impulse within the proton-neutron mass energy balance, the *Strong Interrelation.*
There are neither flying *photons* nor flying *gluons* in the elementary world as the *Standard Model* proposes it. Instead the *Quantum Membrane,* the loaded state of the *Quantum System of Reference* transfers all impacts.
Quantum does not take and does not give off energy and mass: *quantum* itself is the impact, propagating in any direction.

The acceptance of the definitions of the *Quantum System of Reference* and the *Quantum Membrane* would save modern physical science from the ideas of the definition of *Dark Matter* and the *Dark Energy.*

"Particles" have been found, their impacts could be measured in high speed collision experiments because the time system of elementary processes has been *slowed down* by the increased speed of the motion; giving this way chance to measure their increased intensity in our less intensive time system.

A1.1 $dt_i = \dfrac{dt_e}{\sqrt{1-(v^2/c^2)}}$;   and $dt_e = dt_i\sqrt{1-\dfrac{v^2}{c}}$ the time formula

In A1.1 dt_i means the time system of the electron process of the experiment, in motion close to c the actual speed of quantum communication; dt_e is our time system of the experiment on the surface of the *Earth*.

$dt_e << dt_i$ - for this reason (because if the increase of the intensity) elementary events might be measured as particles indeed

Ref. S.6   With reference to Section 6, *quarks* represent certain internal statuses of the infinite cycle of proton-electron-neutron processes. Proton process is with *top-charm-up* process line dominance, neutron process is with *down-strange-bottom* process line dominance.

Down-strange-bottom-top-charm-up-…-down-strange   …processes are measured in line with the intensities of their mass effect: the less the intensity of the impact is, the longer is the chance to measure the event. Therefore there is no surprise that the less intensive processes (like *down* and *up*) are measured more often. They are measurable much longer than others with higher intensity.

The same intensity principle relates to the measurement of leptons, the *electron-muon-tau* process chain as well, each of which is of the same electron function of the *Weak Interrelation*. Just they happen with different intensities, therefore with different options/chances for being detected and measured. *Electron* process is with less value of intensity, therefore mainly measured as drive of the neutron process of the *Strong Interrelation*.

This relates in similar way to *neutrinos a*s well.
Neutrinos as processes belong to all three electron functions: they are the indicators of the deviation of the mass-energy balance of proton-electron-neutron processes at electron process function level.
The *Strong Interrelation* is the one, which generates electron functions (and neutrino's), as the electron function follows the proton process.

Standard Model defines W^+, W^- and Z^o bosons as intermediate vector bosons of the *Weak Force*. Instead both *W* and the *Z* bosons are processes of the mass/energy - energy/mass transformation of the *Strong Interrelation* between internal and external quark processes of protons and neutrons.
We can call them as intermediating vector bosons, but the meaning should be clear.

In the case of *W-s* and *Z* the reason and the effect are different.

W and *Z* bosons are not about electron function and are not *neutrinos;* rather they are the elementary tools of the operating *Strong Interrelation*, representing the energy and mass transfer and mass-energy balance between proton and neutron processes.

This explanation correlates very well with the findings that the half lifetime of both these "bosons" is about $3x10^{-25}$ sec. This is the impact-response of the contiously working *Strong Interrelation* when a disruption happens within the mass-energy balance of the protons and neutron processes, result of high speed collision.

W bosons as processes are acting between hadrons, and are directed from proton to neutron and from anti-neutron to anti-proton.

Z is acting at internal elementary balance level of hadrons and anti-hadrons, and always directed to the direction of the decreasing mass-energy gradient.

Proton process is sphere symmetrical expanding acceleration from $\lim v = 0$ to $i = \lim a\Delta t = c$	With *entropy* and this way *quantum* generating inflexion point between the two.	A1.2
Neutron process is sphere symmetrical accelerating collapse from $i = \lim a\Delta t = c$ to $\lim v = 0$		A1.3

Blue shift impact of the electron function (similar to *bremsstrahlung radiation* in conventional terms) *is* the basic of the *Weak Interrelation*. This is in fact a *blue shift* impact to the *Quantum System of Reference* made by *electron, muon* or *tau* processes. Each of them is representing the same electron process *blue shift* function, sphere symmetrical expanding acceleration processes for infinite time.

The definition of *particles* and *anti-particles* in conventional understanding of quantum physics is missing the main point:
The direction of the gradient of the change determines the "normal" and the "anti-" statuses.

Findings

Higgs boson is not a particle rather a *quantum impact.*
Higgs boson is the quantum impact of the disrupted and newly establishing *Strong Interrelation*, result of the conflict of the colliding or quasi colliding processes.
➤
The mass-energy balance between quark processes of proton and neutron functions of elements is continuous and permanent. If this balance has been destroyed by the collision within LHC, it is re-establishing immediately, since no proton function can exist without neutron function.

This re-establishing energy communication is the impact to the *Quantum Membrane* of the colliding tunnel. This impact is obviously measured via γ γ jets, as they are quantum impacts indeed.

> [γ and γ are not flying photons as supposed to be by the measurement, but massive quantum impacts, transferred to the detectors of the measurement by the *Quantum Membrane* of LHC. There are no flying particles. The quantum impact is propagating within the *Quantum System of Reference.*]

The disruption of the balance needs at least two interfering and colliding "particles". Therefore the two *gamma* jets have been generated as signals of the two destructions.

➢

The quantum impact of the *Strong Interrelation* is working in three different ways:
  (a) It is generated as quantum impact energy cover, acting from the quark processes of the proton process to the quark processes of the neutron process. This energy transfer process in conventional terms has been addressed as W^+ boson force.
  (b) It is generated as quantum impact energy cover, acting from the anti-quark processes of the anti-neutron process to the anti-quark processes of the anti-proton process. This process in conventional terms has been named as W^- boson force.
  (c) It is generated within the internal processes of hadrons, acting from the (t-c-u) quark process chain to (d-s-b) in protons and neutrons; and from (b-s-d) quark process chain to the (u-c-t) in anti-neutrons and anti-protons. This quantum energy cover in conventional terms is identified as Z boson force.

> [Contrary to the conventional understanding W^+, W^- and Z bosons are tools of the mass-energy balance and transfer of the *Strong Interrelation*.]

➢

The quantum energy cover from the proton process of the *Strong Interrelation* needs neutron collapse demand. The neutron collapse however works only if it is driven. The drives are the processes of the electron function, leptons of different intensities (*electron, muon, tau*).

As keys to the re-establishment of the *Strong Interrelation*, leptons are acting in both destroyed elementary structures of the particles in collision, driving neutron collapse the most intensive way. Leptons drive the (d-s-b) quark process chain, rehabilitating with that the quantum energy cover of the elements. In anti-quark processes the (b-c-d) chain is the one which is driven but with decreasing energy accumulation gradient.

As leptons are products of the proton process, the internal elementary balance may need corrections. The appearing forms of the correction are the measured *neutrinos.*

> [Contrary to the conventional understanding, leptons (the *Weak Interrelation*) are the process drives, initiating W^+, W^- and Z boson, the tools of the *Strong Interrelation.*]

➢

For providing an "emergency" (as quick and intensive as possible) drive for re-establishing the natural elementary balance of the *Strong Interrelation* of the destroyed by the collision hadron processes, *tau* process is the most intensive electron function. And the most intensive collapse to be driven in quark processes ends by *bottom* quark process.

Therefore there is no surprise their impacts have been found at the detectors. In anti-quark circumstances, quark process chain ending by *top* quark as it is the most energy intensive process.

Conclusion

Higgs boson is a process, the integrated effect of the *Strong Interrelation.*

The two *gamma* signals, the *4 leptons*, the *WW* and *ZZ* findings, the *tau-tau* and *bottom-bottom* quark processes, all in parallel prove it. (The only missing signal is the *top-top* processes, the other most intensive process of the anti-quark process chain of the *Strong Interrelation.*)

The sum of the absolute values generated by *W* and *Z* bosons – within processes and anti-processes going on in parallel – nearly gives the value of the *Higgs boson,* doubled in the equation – as acting for the processes and the anti-processes as well

$$\left|W^+\right| + \left|W^-\right| + \left|Z\right| \cong \left|2H\right| \qquad\qquad \text{A1.4}$$

Absolute values shall be taken as normal. Anti-processes relate only to the gradient of the change. The energy itself is provided.

There is an important comment to the conclusions.
The massive magnetic field of the electron function from the huge magnets for acceleration results in increased *blue shift* conflict in the LCH channel. The *blue shift* conflict is an overload to the *Quantum Membrane* and results in increased *c*, in increased speed of quantum communication.

Ref.   With reference to Section 6, the following diagram illustrates the chain of
S.6   quark processes within proton / neutron and anti-proton / anti-neutron
        relations.

process side with normal gradient

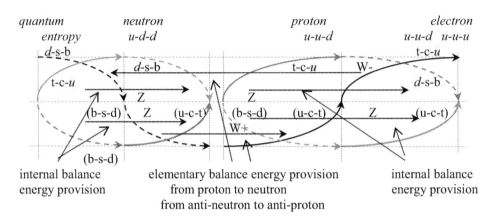

anti-process side with gradient of opposite direction

Fig.
A1.1   Fig. A1.1

DECOMIX
A mix for rehabilitation of isotopes

Basics

The *Decomix* is a mixture of minerals for speeding up the rehabilitation process of isotopes. The speeding up means shortening the half-life time cycles and this way decreasing the total rehabilitation period.

The format of the *Decomix* is powder.

The following elements are the components of the *Decomix* powder. All of them are used in mineral formats, specified in the technology section:

H	Hydrogen
O	Oxygen
N	Nitrogen
C	Carbon
S	Sulphur
Ca	Calcium
Si	Silicon
Mg	Magnesium
Al	Aluminium
K	Potassium
Cl	Chlorine
Na	Sodium
Ti	Titan
Fe	Iron

The elementary components of the *Decomix* communicate with the isotopes at quantum level.

(*Decomix* powder was developed for the decontamination of the contaminated soil around the Fukushima Nuclear Power Plant, consequence of *flooding* and *Earthquake.*)

Decomix can be used for the decontamination of soil, walls, constructions, tools, spare parts, wear etc. contaminated by all kind of isotopes.

In the case of mixing the *Decomix* powder into concrete structures, it provides decontaminating capabilities and characteristics to the construction as well.

Decomix can also be used in nuclear waste treatment technology, either as instrument of the treatment or storage or both.

As result of the rehabilitation, isotopes return into balanced elementary status, while the donor elements of the *Decomix* become isotopes.

The benefit of the rehabilitation is that the generating new isotopes of the donor elements of the *Decomix* have very short half-life time periods:

seconds and *minutes*.

With the half-life time of the isotopes shortened, contaminated subjects (territory, tools, spare parts, wear etc.) become *clean* and *decontaminated*.

In order to decontaminate the *Fukushima* site, the soil shall be sprinkled by *Decomix* powder. The expected full rehabilitation of the isotopes, the total duration of the "cleaning process" is not exceeding 100 days.

Ref.
S.5

The mixing process

The mixing of the components of the *Decomix* powder needs specific technology.
The principle of the mixing is to keep the energy/mass balance between the elementary processes of the components permanently acting.
Elements of the mix with electron process *blue shift* surplus are: *O, H, N, C, S, Ca, Si.*
The other part, with elements *Mg, Al, K, Na, Cl, Ti, Fe* all have increased speed of quantum communication with significant electron process *blue shift* impact potential, but their electron process *blue shift* potential is with deficit.

The intensive quantum communication between the two parts of the mix and the isotope (the rehabilitation itself) starts under the effect of the so called *Initiator*, a specific composition of elements from both groups. It drives the process for establishing the balance.

Permanent temperature measurement during the mixing process is the best indicator of the correctness of the technology. Temperature measurement guaranties the control of the full process. The temperature of the mix, during the preparation shall not be increasing. The mixing technology is structured the way that temperature of the mix can be kept quasi constant.

During its preparation the *Decomix* is also in quantum communication with the environment.

Direct water connection during the mixing process shall be avoided, as water takes away the driving electron *blue shift* surplus of the mix. (The effect of the water however during the rehabilitation process itself is different. Water is with energy surplus, just this surplus is of different potential during the mixing process and during the rehabilitation.)

The prepared *Decomix* is stable, efficient and safe. The powder ensures the balance and the necessary working effect through its conflicting elementary relations.

The precise description of the composition and the technology is patented. The proportions of minerals with elementary components shall be measured before mixing and the proportions shall be kept accordingly.
All mineral components of the mix are of powder format precise and exact size.

The use of minerals for the composition of the mix has its specific reason:
Elements within minerals have increased quantum communication. The speed of the quantum communication of the elements in the minerals is more than in their "clean" laboratory status.

> This also means that mixing these elements in their laboratory clean status rather than in the format of their minerals might not provide rehabilitation and decontamination features to the mix.

Assessment
of the results of the Fukushima experiment of the *Decomix* mixed with *CNH*, the extra clean *Carbon*

(for making the experiment the *Decomix* powder was mixed with extra clean Carbon powder, named: *CNH*)

The measured results well demonstrate the decontamination effect of the *CNH-Decomix*. The half-life of isotopes with many long years can be reduced to couple of months. The use of minerals for the composition of the mix has its specific reason:

Radiation data of the samples treated by *CNH-Decomix* vary in the diapason of	*15-24 cpm*
while the measured activity of the contaminated Fukushima soil is of	*24-49 cpm*

With reference to the measured data there could be questions, relating to the differences in the measurements *from in front* and *from above*.

Explanation is given here below.

CNH-Decomix provides
              – as the key point of the decontamination/rehabilitation effect –
the missing electron process *blue shift* drive and proton process cover to the neutron processes of the damaged elementary cycles of the isotopes of the Fukushima soil.
As result, the energy/mass balance of the damaged elementary processes is rehabilitating and the measured contamination of the Fukushima soil is getting less and less.

In the case of the measurement *from above* the impact of *gravitation* shall be taken into account.

While the rehabilitation of the soil treated by *CNH-Decomix* is going on, *gravitation* modifies *the measured blue shift* need (-*beta radiation*) of the damaged elements of the soil.

The principle of the activity measurement is detection of the change if any in the generated by the measurement device electromagnetic field.

In the case of the measurement of the specimen without *CNH-Decomix* from above, the *beta(-)* isotopes of the Fukushima soil sample take portion of the generated by the measurement device *blue shift* impact of the electromagnetic field of the measurement.

A2.1   If the intensity of the generated by the measurement device *blue shift* impact is $e_{device} = e_d$ and *beta(-)* isotopes are taking portion $\Delta e_{isotopes} = \Delta e_i$ of this impact the remaining intensity of the *blue shift* impact of the device is:

A2.2 $e_{remaining}^{isotope} = e_r^i$.   It is $e_r^i = e_d - \Delta e_i$

As *gravitation* is acting, this remaining *blue shift* intensity will be *blue shifted* by *gravitation* and measured by the device. It is:

A2.3
$$e_{measured}^{isotope} = e_m^i = e_r^i \left[1 + \left(1 - \sqrt{1 - \frac{(g\Delta t)^2}{c^2}} \right) \right]$$

And this *blue shifted* by *gravitation* impact will be measured finally by the measurement device as result.

In the case of the measurement when isotopes have been treated by *CNH-Decomix*, the input of the measurement is different, since *CNH-Decomix* is treating the isotopes, providing the electron process *blue shift* need and proton process cover to their damaged elementary processes.

The measurement device generates the e_d intensity electromagnetic field the same way but in this case the *blue shift* need partially or in full is coming from the elements of the *CNH-Decomix*, since they are becoming damaged as consequence of their elementary support to the isotopes.

A2.4   The *blue shift* need of the elements of the *CNH-Decomix* is: $\Delta e_{Decomix} = \Delta e_x$.

The remaining intensity of the *blue shift* impact of the device will be:

A2.5
$$e_r^x = e_d - \Delta e_x$$

The *blue shifted* by *gravitation* this *blue shift* impact, to be measured by the device is:

A2.6
$$e_{measured}^{decomix} = e_m^x = e_r^x \left[1 + \left(1 - \sqrt{1 - \frac{(g\Delta t)^2}{c^2}} \right) \right]$$

The intensity of the quantum communication of heavy isotopes like *Co, Sr, Cs* and others is high, 10-15% higher than that of the elements within the *CNH-Decomix*. Therefore the demand in electron process *blue shift* of the isotopes is of higher value than the electron *blue shift* need of the damaged elements of *CNH-Decomix*: $\Delta e_i > \Delta e_x$

This means, the remaining impact of the device in the case of the treated Fukushima soil specimen is less than that in the case of *CNH-Decomix* $e_r^i < e_r^x$ and   A2.7

the measured *blue shift* impact of *gravitation* to the electromagnetic field of the measurement device without the treatment by *CNH-Decomix* is also less than with treatment: $e_m^i < e_m^x$ A2.8

In the case of the measurement *from in front* gravitation effect is excluded and the measured data give the real data.

As consequence of the rehabilitation, there will be isotopes formulating within the *CNH-Decomix*, but those are with short range of half-life with seconds and minutes.

The measured data also prove that the rehabilitation effect is coming from the decontamination mix – *Decomix*. *CNH* is not adding rehabilitation impact to the composition.

> Comment: There was no reference measurement in the case of "from above". The "from in front" reference was used. Otherwise the new reference would be different, equal to the energy intensity of the measurement device.

Results

Ref Table A.2

The measured data are given in Table A.2 below.

The quantum speed of these isotopes is increased and is taken as c_i.

The speed of quantum communication of the *blue shift* impact from the device is c_d.

> The external *blue shift* impact of the measurement device is without proton cover and is only for the purpose of making the measurement.
> Electron processes are of constant time systems, therefore they are communicating.

As isotopes are heavy meatels: $c_i > c_d$

This way, there are *Quantum Membranes*

at one side with c_i - plus neutron dominancy; and c_d on the other.

> Heavy metals are with *blue shift* deficit and therefore with neutron process dominance of high intensity.

Measured data   from above ↑

                                          ⌐———→ from in front

From in front

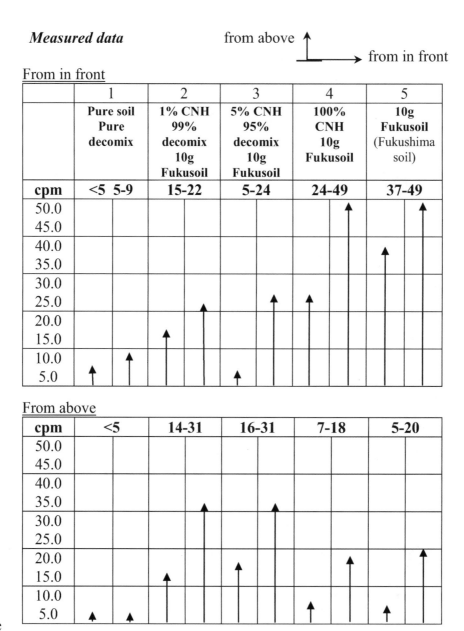

	1		2	3	4	5
	Pure soil Pure decomix		**1% CNH 99% decomix 10g Fukusoil**	**5% CNH 95% decomix 10g Fukusoil**	**100% CNH 10g Fukusoil**	**10g Fukusoil (Fukushima soil)**
cpm	**<5**	**5-9**	**15-22**	**5-24**	**24-49**	**37-49**
50.0 45.0						
40.0 35.0						
30.0 25.0						
20.0 15.0						
10.0 5.0						

From above

cpm	**<5**		**14-31**	**16-31**	**7-18**	**5-20**
50.0 45.0						
40.0 35.0						
30.0 25.0						
20.0 15.0						
10.0 5.0						

Table
A.2

Table A.2

While the external *blue shift* impact of the measurement does not provide proton cover, quantum systems need balance. Therefore elements with increased speed of quantum communication also are very keen to take external electron process *blue shift* impact from the *Quantum Membrane* generated by the measurement device even without proton cover.

Increased neutron process dominance makes elements less stable as the deviation between the intensities of the proton and neutron processes makes these elements soft. [This *softness* is different than the one close to liquid state, which is from electron process surplus.]

Soft elementary structures with decreasing quantum speed are in fact hardening, as the intensity of the electron process drive, with stable and constant proton/neutron intensity relations is decreasing.

(In other words: The neutron process dominance is "eating off" external *blue shift* impacts while proton processes go behind, because of their less intensity.)

The decreasing quantum speed is consequence of the quantum system with speed of integrated quantum communication of less value.

Electron processes are of the same time system:

$$n_{xx}\frac{dmc_{xx}^2}{dt_i}\left(1-\sqrt{1-\frac{(c_{xx}-i_{xx})^2}{c_{xx}^2}}\right)+n\frac{dmc^2}{dt_i}\left(1-\sqrt{1-\frac{(c-i)^2}{c^2}}\right)=$$

$$=n_x\frac{dmc_x^2}{dt_i}\left(1-\sqrt{1-\frac{(c_x-i_x)^2}{c_x^2}}\right)$$

A2.9

$$\text{and } c_{xx}>c_x>c \qquad\qquad\qquad \text{A2.10}$$

As the quantum speed of the isotopes is higher than the integrated quantum speed of the *CNH-Decomix*, the intensity of the *blue shift* impact taken from the *Quantum Membrane* of the measurement device by the not treated isotopes is more than taken by the samples treated by *CNH-Decomix*.

The measurement device is getting back less value of *blue shifted blue shift* impact from isotope to be measured.

The measurement device is getting back *blue shifted* by *gravitation blue shift* impact in more value from the samples treated by *CNH-Decomix*.

A3   **The quantum energy of rain drops**

The energy of rain can be measured.

With reference to Section 9, the natural quantum speed values of the *Oxygen* and the *Hydrogen*, components of water, are less than the quantum speed of *gravitation*. The "forced" by *gravitation* increased quantum speed and the electron process *blue shift* surplus of these elements make the air of *Earth* atmosphere vapour, steamy, fogy, cloudy.

In other words:

The *Quantum Membrane* above *Earth* surface contains water processes with conflicting *blue shift* impacts. The working *blue shift* conflict keeps water in vapour, steam and fog or cloud states. If, as result of weather conditions, the level of the conflict is less, water takes its natural liquid state. Drops are formulating, as the water process in less conflict occupies less quantum space.

Rain is coming with pressure and temperature decrease (in conventional understanding) – with the decrease of the conflict and the intensity of the *Quantum Membrane*.

This is the physical status when water drops become being in" free fall".

Water drops are no more part the *Quantum Membrane* above the *Earth* surface in conflicting quasi elementary process format. The slowdown of water drops in free fall by *g* relative to *gravitation* increases. The speed of the free fall of the drops relative to the *Earth* surface is increasing.

Each water drop in free fall is in "mechanical" conflict with all other elementary processes of the air in *gravitation*. This mechanical conflict helps to further disperse drops and make their size as less as possible.

Water drops gather speed and cause massive break of *Earth* magnetic lines. The impact of the interruption of the *Strong Interrelation* of magnetic lines generates *blue shift* conflict: rain is bringing energy to the *Earth* surface.

Rain is energy.

With reference to Section 11, *Earth* surface is with proton process dominance. Water drops are bringing electron process *blue shift* surplus and intensify elementary processes within the soil.

Vegetation is grateful for the gift.

The energy content of water drops can be measured by the net of very thin *Cupper* wires from electricity cables, mounted at a wooden structure.

With reference to Figure A3 below, rain drops with increased intensity of their electron process *blue shift* impact, increase the intensity of the internal electron process *blue shift* impact of the *Copper* element of the net of wires in two levels. The *blue shift* conflict generates surplus within the wires, which can be measured and taken away.

As result of the experiment with the drops of a normal summer rain, the generated electron process *blue shift* surplus, the measured energy, the generated electricity within the *Copper* wires was up *0.19 mA.*

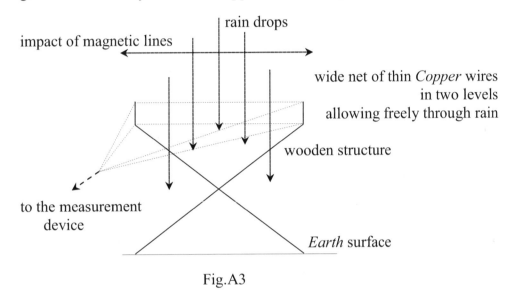

Fig.A3

Fig.
A3

Rain drops "falling" through the wide net generate electricity flow within the wires.

Here the electricity generation can be divided into two parts:
(1) the break of magnetic lines by water drops generates electron process *blue shift* surplus and conflict within the drops;
(2) the constant impact of rain drops with increased electron process *blue shift* intensity impact generates electron process conflict and surplus within the wire.

Measuring electricity within the net of wires takes time, as rain impact has to generate sufficient measurable effect. The advantage of the experiment with rain drops is the electron process *blue shift* surplus of the water.